Studies in Logic

Volume 64

Logic of Questions in the Wild

Inferential Erotetic Logic in Information Seeking Dialogue Modelling

Studies in Logic Series Editor
Dov Gabbay   dov.gabbay@kcl.ac.uk

Logic of Questions in the Wild

Inferential Erotetic Logic in Information Seeking Dialogue Modelling

Paweł Łupkowski

ISBN 978-1-84890-216-9

College Publications
Scientific Director: Dov Gabbay
Managing Director: Jane Spurr

http://www.collegepublications.co.uk

Printed by Lightning Source, Milton Keynes, UK

To my Daughter

Contents

Introduction

A: Hey, what are you doing?

B: Why, are you writing a book or something?

Such a response was something I encountered quite often during my childhood conversations. I have to admit that back then I was more irritated by this type of response than interested in it. The natural expectation when someone asks a question is a declarative information of some sort, not a question. Today a question given as a response to a question, is something that I think is interesting to the extent of writing a book about it.

The main aim of this book is to study the nature of dependency of questions appearing in the context of natural language dialogues. This means that I will focus on erotetic[1] inferences whose premises and/or conclusion is a question, and on the issue of the criteria of validity of these inferences. I will focus especially on situations where the initial question is processed and an auxiliary question is asked in order to facilitate the search for an answer to the initial one.

I am interested in dialogues, where agents have a common goal and cooperate in order to achieve this goal. The dialogue types analysed in this book will belong to the following types: information seeking dialogues and tutorial dialogues. This type of choice allows us to investigate the questioning strategies used by agents involved in a conversation.

As a point of departure this book makes use of A. Wiśniewski's Inferential Erotetic Logic (hereafter IEL, see Wiśniewski 1995, 2013b) and its tools, namely erotetic implication and erotetic search scenarios. These tools are used to analyse the data retrieved from the language corpora and question processing research. The choice of this particular logical framework is natural for the aims of this book as IEL is itself focused on inferences with questions involved.

---

[1] The term 'erotetic' stems from Greek 'erotema' meaning question. The logic of questions is sometimes labelled erotetic logic (see Wiśniewski 2013b, p. 1).

(More reasons for the choice of this particular logical framework are given in detail in Chapter 2.) As such it offers a semantical analysis of such inferences and—what is especially important—proposes certain criteria of their validity.

At this point of the introduction a reader may ask the question: how might logic be used to study natural language dialogues? How will normative and descriptive perspectives be merged together? Let me start here with explaining the title of this book. It refers to Chapter 5 of Stenning's and van Lambalgen's 2008 *Human reasoning and cognitive science* entitled "From the Laboratory to the Wild and Back Again". Not only the title, but also my usage of logical tools are inspired by this book.

The point I am trying to make here is not that logical concepts are the ultimate explanation of linguistic phenomena (i.e. that people are or should process questions according to IEL). My approach here is rather that logic provides a very useful normative yardstick to study, describe and analyse these phenomena (see Stenning and Van Lambalgen, 2008, p. 130). Such an approach opens up new perspectives for logical and empirical research.

> Nowadays we are witnessing a 'practical', or cognitive, turn in logic (...) It claims that logic has much to say about actual reasoning and argumentation. Moreover, the high standards of logical inquiry that we owe to Peano, Frege, Skolem, Tarski and others offer a new quality in research on reasoning and argumentation. (Urbański, 2011, p. 2)

Logic may be, and is, successfully applied within research concerning actual human reasoning as reviewed and discussed in (Urbański, 2011) and (Sedlár and Šefránek, 2014). Recent examples of this type of approach are int. al.:

- (Gierasimczuk et al., 2013)—logical model of deductive reasoning tested *via* the Deductive Mastermind online game;
- (Zai et al., 2015)—dataset of human syllogistic inference used to test the integrating models of Natural Logic and Mental Logic;
- (Noelle et al., 2015)—presenting an ERP study of bare numerals for possible readings;
- (Szabolcsi, 2015)—presenting a study of semantics of quantifier particles which uses inquisitive semantics framework as the main tool.

As I try to show in this book IEL offers convenient tools for modelling natural language phenomena and for their better understanding. Using these tools I will consider the issue of motivation for certain moves in dialogue. It also allows for implementations of such models (e.g. in cooperative systems and dialogue systems—cf. Łupkowski and Leszczyńska-Jasion 2014). What is more, logic of questions provides us with a normative framework useful for designing empirical research based on erotetic reasoning (see e.g. Erotetic Reasoning Test (Urbański et al., 2014), QuestGen game (Łupkowski, 2011b)).

On the other hand, empirical data (like that retrieved from language corpora) allows for *better tailored logical concepts*. We can compare different approaches to questions' dependency, e.g. compliance, erotetic implication, topicality (all of them claiming that they preserve the intuitions from natural language). We can also use the empirical data to modify and improve e-scenarios, e.g.: by introducing the 4-valued logic (Belnap 1977, Szałas 2013) which will allow for inconsistencies to be dealt with, or by using modified erotetic implication (weak e-implication (Urbański et al., 2014), falsificationist e-implication (Grobler, 2012), epistemic e-implication (Peliš and Majer, 2011)).

One technical remark is in order here. By an *answer* I will understand a direct answer (see page 24) and the term *response* will be used in a more general sense of a verbal reaction to a question.

This book consists of five chapters and is structured as described below.

Chapter 1. Question-driven dialogues. How can questions govern a conversation?

The first chapter reviews question rising mechanisms and their role in dialogue management. Two groups of approaches are presented: linguistically oriented and logically oriented. For the first group the TTR/KoS (Ginzburg, 2012) approach is presented in detail and also the ideas of question-driven discourse proposed by van Kuppevelt (1995) and Graesser et al. (1992) are discussed. For the second group the presentation will cover: the notion of compliance in inquisitive semantics (Groenendijk and Roelofsen, 2011), (Wiśniewski and Leszczyńska-Jasion, 2015); epistemic logic of questions and public announcements (Peliš and Majer, 2010, 2011), (Švarný et al., 2013); and a framework for research agendas rooted in Interrogative Games (Genot, 2009a, 2010).

Chapter 2. Questions and Inferential Erotetic Logic

The second chapter introduces the basic concepts of IEL. It is a logic that focuses on inferences involving questions (see page 25), and which provides criteria of validity of such inferences. IEL gives a very useful and natural framework for analyses of the questioning process. I will present how a question is understood in IEL, define basic concepts as: a safe question, an erotetic implication, or an erotetic scenario. Also the formal languages to be used in the following chapters will be introduced here.

Chapter 3. Dependency of questions in natural language dialogues

The aim of the third chapter is to answer the question: How does natural language data relate to the picture presented in Chapter 2? In order to achieve this I will analyse the data retrieved from The Basic Electricity and Electronics Corpus (BEE) (Rosé et al., 1999) focusing on: simple yes/no vs. open questions; patterns of transitions between questions and answering patterns. The first section of this chapter contains modelling concerning the dependency of questions retrieved from BEE and The British National Corpus (BNC). I will focus on the issue of how erotetic implication may serve as a rationale for certain dialogue moves. The second section covers the analysis of broader parts of dialogues with the aim of analysing the perspective and strategies used by interrogators. The language data retrieved from BEE will be used here.

Chapter 4. Dialogic logic and dynamisation of e-scenarios execution

The fourth chapter brings a dynamic view into the picture. Here, I will present a dialogic logic approach to the dynamisation of erotetic search scenario execution. The outcome system allows for the modelling of the dynamic interrogative problem solving with the underlying IEL concepts of erotetic inferences validity. The first section introduces the system for two party dialogue games, while the second one extends this system to its use in a multi-agent environment.

Chapter 5. Cooperation and questioning

The last chapter is aimed at the application of the previous considerations into the field of cooperative answering for databases and information systems. Generally, we may say that the systems in question are capable of providing direct answers, i.e. information directly requested by the user's question. However, we would expect a cooperative response—which is understood as an informative reaction (non-misleading and useful) to the user's question. Certain cooperative answering phenomena are modelled within the framework of e-scenarios. E-scenarios are used here to generate cooperative responses in cases when: the answer to a question is negative or there is no answer available in a database. These techniques are supplemented with the idea of cooperative question-responses as an extension of traditional cooperative behaviours. I present a procedure to generate question-responses based on question dependency and erotetic search scenarios. The last section of this chapter introduces the dialogue logic system for cooperative answering in the style presented in Chapter 4.

The dependency diagram of chapters of this book is presented below.

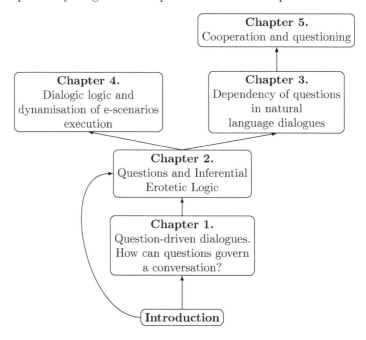

This book presents certain logical concepts placed in the context of vivid and dynamic natural language in dialogue. Such an endeavour requires unavoidable simplifications and restrictions laced on both sides of the story. I hope, however, that despite this the final effect of this specific tailoring will be interesting for the dialogue research community and for logicians as well. The underlying idea would be a kind of a loop-input for logical concepts concerning reasoning (not all of them certainly) leading from logic through the empirical domain as a form of a testing field and back to logic again.

Acknowledgements I would like to thank Mariusz Urbański for his helpful feedback and comments on a draft of this book.
Work on this book was supported by funds provided by the National Science Centre, Poland (DEC-2012/04/A/HS1/00715).

Chapter 1

Question-driven dialogues. How can questions govern a conversation?

This chapter consists of an overview of frameworks and approaches concerning the role of questions in interaction. This issue is getting more and more attention among researchers. As Ginzburg points out:

> there is still vigorous discussion concerning the issue of what questions are and how best to characterize them. Nonetheless, there seems to be an emerging consensus about the need of adopting a dialogical perspective within which such characterization should take place. (Ginzburg, 2011, p. 1143)

Roughly we can speak about two groups of approaches here. One is driven by empirical conversational phenomena with the aim of providing a detailed linguistic analysis of them (see Ginzburg, 2011, p. 1138). The other takes logic as its starting point and aims at accommodating linguistic phenomena. In what follows I will present chosen ideas from both groups.

What is characteristic of all the approaches presented here is that they try to grasp the idea behind the relevance of questions when used in a dialogue. Why is a question asked, and when it is reasonable/profitable to ask a question? How can a question be answered? Can a question be answered with a question? There are certain topics and intuitions that they all share. What differentiates them, are the tools they use. By choosing approaches presented here I want to introduce these intuitions and solutions, that will be later grasped on and unified under the framework of IEL. In what follows I present a given approach and then discuss how the results or techniques will be used for considerations in this book.

I should stress here, that the presented overview is far from being complete. At the end of the chapter I refer to papers offering broader reviews concerning questions and dialogue topics.

1.1. Linguistically oriented approaches

When it comes to linguistically oriented approaches I start by presenting the KoS framework which is based on Type Theory with Records. KoS allows us to model many aspects of dialogical relevance in an intuitive way. In what follows I will also describe the ideas of question-driven discourse proposed by van Kuppevelt (1995) and Graesser et al. (1992).

1.1.1. TTR and KoS

An extensive framework that allows for analysing questions in a dialogical context is provided by KoS.[1] KoS is a framework formulated using Type Theory with Records (TTR) (Cooper, 2005, 2012; Cooper and Ginzburg, 2015). The framework aims at providing a theory of broadly understood relevance that goes beyond 'semantic answerhood' (see Ginzburg (2011, p. 1139) and Ginzburg (2010)).

KoS allows for an explanation of the following dialogue relevance types:

Q(uestion)-specificity   this includes both answerhood and some sort of dependence or entailment relation between questions. Q-specificity will be discussed in detail further on.

Metadiscursive relevance   covers interactions concerning what should or should not be discussed at a given point in a conversation, as in the following example (Ginzburg, 2011, p. 1139):

A: Are you voting for Tory?
B: I don't want to talk about this.

Genre-based relevance   the basic intuition here is that a move can be made if it relates to the current activity, which is exemplified by the following interaction at a train station (Ginzburg, 2010, p. 124):

A: How can I help you?
B: A second class return ticket to Darlington, leaving this afternoon.

B's relevant response in the example is not about A's question but it relates to the current activity (i.e. buying a train ticket).

Metacommunicative relevance   a notion that underwrites clarification interaction. Consider (Ginzburg, 2011, p. 1139):

---

[1] KoS is a toponym—the name of an island in the Dodecanese archipelago—bearing a loose connection to *conversation oriented semantics* (Ginzburg, 2012, p. 2). KoS is used as a formal underpinning for the information state approach to dialogue management (Larsson and Traum, 2003) and dialogue systems such as GoDiS (Larsson, 2002) and CLARIE (Purver, 2006).

A: Are you voting for Tory?

B: Who do you mean 'Tory'?

In what follows I will focus only on the q-specificity relevance type.

KoS analysis is formulated at a level of information states, one per conversational participant. Each information state consists of a dialogue gameboard and a private part. An example of these kinds of information states is given in below.

$$\text{TotalInformationState (TIS)} =_{def} \begin{bmatrix} \text{dialoguegameboard : DGBType} \\ \text{private : Private} \end{bmatrix}$$

In what follows I will focus on the dialogue gameboard part of TIS. The dialogue gameboard (DGB) represents information that arises from publicised interaction. DGB type is represented by the following structure:

$$\text{DGBType} =_{def} \begin{bmatrix} \text{spkr: Ind} \\ \text{addr: Ind} \\ \text{utt-time : Time} \\ \text{c-utt : addressing(spkr,addr,utt-time)} \\ \text{Facts : Set(Proposition)} \\ \text{Pending : list(loc(utionary)prop(osition))} \\ \text{Moves : list(loc(utionary)prop(osition))} \\ \text{QUD : poset(Question)} \end{bmatrix}$$

Fields of DGB allow for tracking an agent's record in the interaction. The fields *spkr* and *addr* track turn ownership. Conversationally shared assumptions are represented in *Facts*. *Pending* and *Moves* represent, respectively, moves that are in the process of being grounded and moves that have been grounded. The *QUD* field tracks (a partially ordered set) of the questions currently under discussion. The following quote illustrates the intuition behind QUD:

> The property of being in QUD that a question can bear is primarily taken to correlate with the fact that the question constitutes a 'live issue': a question that has been *introduced for discussion* at a given point in the conversation and not yet been *downdated*. (Ginzburg, 2012, p. 68)

Dialogue moves are represented as changes in gameboard configuration. Consequently, the basic unit of change is a mapping between dialogue gameboards—such a mapping is called a *conversational rule*. These rules allow for an explanation of how conversational interactions lead to certain conversational states. A conversational rule consists of preconditions (*pre*) and *effect* fields.

$$\begin{bmatrix} \text{pre} & : & \text{PreCondSpec} \\ \text{effects} & : & \text{ChangePrecondSpec} \end{bmatrix}$$

As an illustration of a conversational rule let us consider Ask QUD-incrementation, which specifies how a question becomes established as the maximal element of QUD in DGB.

Ask QUD–incrementation

$$
\left[\begin{array}{l}
\text{pre :} \quad \begin{bmatrix} \text{q :} & \text{Question} \\ \text{LatestMove} = \text{Ask(spkr,addr,q):} & \text{IllocProp} \end{bmatrix} \\[1.5em]
\text{effects :} \quad \begin{bmatrix} \text{qud} = \big\langle \text{q,pre.qud} \big\rangle\text{:} \quad \text{poset(Question)} \end{bmatrix}
\end{array}\right]
$$

The rule says that given a question q asked by the speaker to the addressee in the latest move one can update QUD with q as its maximal element. After a question q becomes the maximal element of QUD an utterance will appear in dialogue as a reaction to this fact.

Let us consider here dialogical moves in which the responder accepts question q as an issue to be resolved. In such a situation the **QSPEC** conversational rule will be in use.

QSPEC

$$
\left[\begin{array}{l}
\text{pre :} \begin{bmatrix} \text{qud} = \big\langle \text{q, Q} \big\rangle\text{: poset(Question)} \end{bmatrix} \\[1em]
\text{effects : TurnUnderspec} \wedge_{merge} \\[0.5em]
\begin{bmatrix} \text{r : Question} \vee \text{Prop} \\ \text{R: IllocRel} \\ \text{LatestMove} = \text{R(spkr,addr,r) : IllocProp} \\ \text{c1 : Qspecific(r,q)} \end{bmatrix}
\end{array}\right]
$$

QSPEC *effects* field states in the first place that the turn is underspecified (i.e. the next dialogical move may be taken by any of the dialogue participants). This allows for capturing cases of self-answering or multiparty dialogue contexts. In what follows the **QSPEC** rule introduces the *q-specific* dialogue move (see Ginzburg, 2012, p. 57).

Definition 1 (q-specific utterance) *An utterance whose content is either a proposition p* About *q or a question q_1 on which q* Depends.

The intuition behind a proposition p being about a question q is that p is at least a partial answer to q. The dependency of questions is explicated as follows (see Ginzburg, 2012, p. 57).

Definition 2 (Question dependence) q_1 depends on q_2 *iff any proposition p such that p* resolves *q_2, also satisfies p entails r such that r is about q_1.*

The intuition behind this notion is that whenever an answer to question q_2 is reached then some information about the answer to the q_1 becomes available.

The following dialogue illustrates this idea:

A: Does anybody want to buy an Amstrad? *<pause>*
B: Are you giving it away?
    [*BNC: KB0, 3343–3344*]
    [*Whether anybody wants to buy an Amstrad depends on whether you are giving
    it away.*]

Dialogue gameboards and conversational rules allow for the expression of the *coherence of a dialogue move*. The very intuition is that a move m_1 is coherent to a dialogue gameboard dgb_0 just in case there is a conversational rule c_1 which maps dgb_0 into dgb_1 and such that dgb_1's LatestMove field value is m_1. For a formal definition of move coherence and its extension to *pairwise-* and *sequential*-move coherence see (Ginzburg, 2012, p. 96).

What is more, the DGB structure allows for the expression of certain conditions when it comes to asking a question in a dialogue. This condition is formulated as the Question Introduction Appropriateness Condition (QIAC):

> a question q_1 can be introduced into QUD by A only if there does not exist a fact τ such that $\tau \in Facts$ and τ resolves q_1 (Ginzburg, 2012, p. 89).

This section contains only a short presentation of certain key concepts of the KoS framework. An interested reader will find a detailed KoS presentation along with the discussion of underlying intuitions and many examples in Ginzburg (2012), Cooper and Ginzburg (2015) and Ginzburg (2016).

When discussing q-specific utterances in the context of KoS, erotetic implication is often pointed out as a concept that grasps the idea of question dependency (see Ginzburg 2010, p. 123, Ginzburg et al. 2014, p. 93, Moradlou and Ginzburg 2014, p. 117). In Chapter 3 of this book I present a detailed analysis of the erotetic implication in this context.

One may notice that since we are considering DGBs, we take into account a kind of public knowledge repository of dialogue participants rather than their individual knowledge (see Ginzburg, 2012, p. 88). In Chapter 4, I present a dialogue logic system rooted in IEL concepts that shares certain intuitions with the KoS framework, such as e.g. tracking turn ownership and analysing a public knowledge repository of dialogue participants.

1.1.2. Topicality as the organising principle in discourse

In what follows I will present two interesting concepts concerning questions in dialogue. The first one, is van Kuppevelt's *topicality* as the general organising principle in discourse. The second one, aims at providing a typology of question-generating mechanisms for dialogues (especially in educational contexts).

What is interesting in the proposal of van Kuppevelt (1995) is that it aims at providing an explanation of a structural coherence of discourse. What is more, not only explicitly formulated questions are considered here but also implicit ones are taken into account. The notion of *topic* is defined in the following way (see van Kuppevelt, 1995, p. 112):

Definition 3 (Topic) *[A] discourse unit **U**—a sentence or a larger part of discourse—has the property of being, in some sense, directed at a selected set*

*of discourse entities (a set of persons, objects, places, times, reasons, conse-
quences, actions, events or some other set), and not diffusely at all discourse
entities that are introduced or implied by* **U**. *This selected set of entities in
focus of attention is what* **U** *is about and is called the* topic *of* **U***.*

The basic assumption of the proposed framework is that questions introduced into a discourse (explicit or implicit) constitute topics. The following
example (see van Kuppevelt, 1995, p. 113) illustrates this assumption:

A: Late yesterday evening I got a lot of telephone calls. [F]
B: Who called you up? [Q_1]
A: John, Peter and Harry called me up. [A_1]

In the presented dialogue a topic (i.e. who called up speaker A) is introduced
by the explicit question Q_1.

For a question (implicit or explicit) to appear in a dialogue a so called *feeder*
is necessary. Feeders may be of a linguistic or non-linguistic (situational, con-
textual) character. A linguistic feeder is a topicless unit of discourse (see van
Kuppevelt, 1995, p. 119). Feeders initiate or re-initiate the process of questioning. In dialogue examples provided in this section feeders are marked by
[F]. Questions asked as the result of the feeder are called *topic-constituting*
questions (see van Kuppevelt, 1995, p. 122).

Definition 4 (Topic-constituting question) *An explicit or implicit question Q_p is a topic-constituting question if it is asked as the result of a set of
preceding utterances which, at the time of questioning, functions as a feeder.*

Sometimes answers provided to questions are not satisfactory. This type of
situation gives rise to a *subquestion* constituting a so called *subtopic*.

As van Kuppevelt (1995, p. 123) puts it, the aim of introducing subquestions is to provide a satisfactory answer to the topic-constituting question (van
Kuppevelt, 1995, page 125):

Definition 5 (Subtopic-constituting subquestion) *An explicit or implicit
question Q_p is a subtopic-constituting subquestion if it is asked as the result of
an unsatisfactory answer A_{p-n} to a preceding question Q_{p-n} with the purpose
of completing A_{p-n} to a satisfactory answer to Q_{p-n}.*

What is particularly interesting is that the notion of an *unsatisfactory* answer is not the absolute one, it is relative to a dialogue participant's background
knowledge and assumptions.

The following example (see van Kuppevelt, 1995, p. 129) presents the idea of
subquestions (Q_2, Q_3 and Q_n) appearing after the unsatisfactory answer (A_1)
to topic-constituting question Q_1:

A: It's sensible for Tom to buy a car now. [F]
B: Why? [Q_1]

A: Buying a car is probably favourable for him now and it won't be bad for his health. [A_1]

B: Why is buying a car probably favourable for him now? [Q_2]

A: Car expenses are expected to decrease. [A_2]

B: For what reason? [Q_3]

A: Gas will become substantially cheaper. [A_3]

B: Why won't a car be bad for his health? [Q_4]

A: He jogs every day. [A_4]

The process of subquestioning is guided by two principles (van Kuppevelt, 1995, p. 129):

1. *Principle of recency* indicating that subquestions are introduced as a result of the most recent unsatisfactory answer to a preceding question.
2. *Dynamic principle of topic termination* stating that if a (sub)question is answered satisfactorily the questioning process associated with it comes to an end.

The intuitions of the topicality framework are well-reflected in the way the questions are decomposed within the IEL approach. This may be noticed in Chapter 2, where the notions of erotetic implication and erotetic search scenario are introduced and in Chapter 3 where natural language dialogues are analysed (especially in Chapter 3.2 where certain topic-constituting questions in the analysed examples are implicit).

1.1.3. Question-generating mechanisms

The so called GPH scheme[2] focuses on the issue of question-generation mechanisms in natural settings (including tutorial dialogue).

Graesser et al. (1992) propose four main question-generation mechanisms: *knowledge deficit* questions, *common ground* questions, *social coordination* questions and *conversation-control* questions. Knowledge deficit questions appear when a questioner detects a lack of knowledge in his or her own knowledge base. These are asked in order to seek out the missing information. Common ground questions, like 'Are we working on the third problem?' or 'Did you mean the independent variable?', are asked to check whether knowledge is shared between dialogue participants. Social coordination questions relate to different roles of dialogue participants, such as in student–teacher conversations. Social coordination questions are requests for permission to perform a certain action or might be treated as an indirect request for the addressee to perform an

---

[2] The name refers to its creators: Graesser, Person and Huber. GPH is presented in detail and discussed in (Graesser et al., 1992) and (Graesser and Person, 1994). Olney et al. (2012) propose a semi automatic model of question-generation of the basis of knowledge base represented as concept maps.

action (e.g., 'Could you graph these numbers?', 'Can we take a break now?'). Conversation-control questions, as indicated by their name, aim at manipulating the flow of a dialogue or the attention of its participants (e.g., 'Can I ask you a question?').

From my perspective in this book, the most interesting is the first group—i.e. knowledge deficit questions. Such information-seeking questions may occur under the following conditions (see Graesser and Person, 1994, p. 112–113):

1. the questioner encounters an obstacle in a plan or problem,
2. the questioner encounters a contradiction,
3. an unusual or anomalous event is observed,
4. there is an obvious gap in the questioner's knowledge base, or
5. when a questioner needs to make a decision among a set of alternatives that are equally likely.

Graesser and Person (1994) provide examples of such knowledge-deficit questions extracted from an actual tutoring session. In (1a) a student realises a gap in her knowledge base and in (1b) the student has spotted a contradiction between his own belief and the tutor's expressed doubt.

(1) a. TUTOR: Cells are also the same thing as groups or called experimental conditions. So those little boxes could be called cells, they could be called groups, they could be called experimental conditions.
         STUDENT: Wait a minute. When you say "boxes" what do you mean?

     b. TUTOR: Is there a main effect for "A"?
         STUDENT: I don't think so.
         TUTOR: You don't think so?
         STUDENT: (Laughs.) Is there one?

What is interesting in the presented approach is that Graesser and Person (1994, p. 108) point out that in an educational context we may notice questions that are not genuine information-seeking questions. Roughly speaking, for a genuine information-seeking question to be asked the questioner should be missing information and believe that the answerer can supply it. When it comes to teacher questions (e.g. in tutorial dialogues) often these requirements might not be met. As Graesser and Person (1994, p. 125) report the results of their study of 27 tutoring protocols, the vast majority of questions asked by tutors in the analysed sample, are in fact common ground questions aimed at an assessment of the student's knowledge.

Situations like these will be discussed in Chapter 3.2 where tutoring dialogues are analysed for the questioner's strategies.

1.2. Logically oriented approaches

This section is devoted to logically oriented approaches to the issue of questions in interaction. Three main focus points may be observed here. Namely (i) concern over the issue, when it is rational to ask a question; (ii) questions as q-specific responses and (iii) a strategic questioning process (questioning agendas).

Firstly, I present the interesting framework of Inquisitive Semantics (INQ). The approach to what is a question is different than that of IEL for example, as we will see, the intuitions of the role of questions in interaction are shared between these frameworks. After INQ, I present two approaches to the issue of belief revision during agents' interaction and the role of questions in this process (in the context of strategy behind questioning and the informative role of asking questions). The first framework is rooted in IEL, while the second stems from Hintikka's Interrogative Games.

1.2.1. Inquisitve Semantics and the notion of compliance

In the framework of inquisitive semantics dependency between questions (viz. q-specificity) is analysed in terms of *compliance*. The intuition behind the notion of compliance is to provide a criterion to 'judge whether a certain conversational move makes a significant contribution to resolving a given issue' (Groenendijk and Roelofsen, 2011, p. 167). If we take two conversational moves: the initiative A and the response B, there are two ways in which B may be compliant with A (cf. Groenendijk and Roelofsen 2011, p. 168):

1. B may partially *resolve* the issue raised by A (answerhood).
2. B may replace the issue raised by A by an easier to answer sub-issue (sub-questionhood).[3]

Here I am be interested only in the case when we are dealing with sub-questionhood. Before I provide a definition of compliance, I first introduce the necessary concepts of INQ, especially the notion of question used in this framework. My main source for this task will be a paper by Wiśniewski and Leszczyńska-Jasion (2015), which presents an exhaustive and rich description and discussion of the most often used system of inquisitive semantics, labelled as InqB.[4]

Firstly (after Wiśniewski and Leszczyńska-Jasion 2015) let us introduce a language $\mathcal{L_P}$. It is a propositional language over a non-empty set of proposi-

---

[3] In inquisitive semantics also combinations of (a) and (b) are also possible, i.e. B may partially resolve the issue raised by A and replace the remaining issue by an easier to answer sub-issue.

[4] For papers presenting various versions of INQ and recent developments see: http://sites.google.com/site/inquisitivesemantics/.

tional variables $\mathcal{P}$, where $\mathcal{P}$ is either finite or countably infinite. The primitive logical constants of the language are: $\bot$ (falsum), $\vee$ (disjunction), $\wedge$ (conjunction), $\rightarrow$ (implication). Well-formed formulas (wffs) of $\mathcal{L}_{\mathcal{P}}$ are defined as usual.

The letters A, B, C, D, are metalanguage variables for wffs of $\mathcal{L}_{\mathcal{P}}$, and the letters X, Y are metalanguage variables for sets of wffs of the language. The letter $\mathbf{p}$ is used below as a metalanguage variable for propositional variables.

$\mathcal{L}_{\mathcal{P}}$ is associated with the set of suitable possible worlds, $\mathcal{W}_{\mathcal{P}}$, being the *model* of $\mathcal{L}_{\mathcal{P}}$. A possible world is identified with indices (that is valuations of $\mathcal{P}$). $\mathcal{W}_{\mathcal{P}}$ is the set of all indices.

A *state* is a subset of $\mathcal{W}_{\mathcal{P}}$ (states are thus sets of possible worlds). I will use the letters σ, τ, γ, to refer to states.

The most important semantic relation between states and wffs is that of *support*. In the case of INQ support, $\succ$, is defined by (see Wiśniewski and Leszczyńska-Jasion, 2015, p. 1590):

Definition 6 (Support) *Let $\sigma \subseteq \mathcal{W}_{\mathcal{P}}$.*

1. $\sigma \succ \mathbf{p}$ *iff for each $w \in \sigma$: $\mathbf{p}$ is true in w,[5]*
2. $\sigma \succ \bot$ *iff $\sigma = \emptyset$,*
3. $\sigma \succ (A \wedge B)$ *iff $\sigma \succ A$ and $\sigma \succ B$,*
4. $\sigma \succ (A \vee B)$ *iff $\sigma \succ A$ or $\sigma \succ B$,*
5. $\sigma \succ (A \rightarrow B)$ *iff for each $\tau \subseteq \sigma$: if $\tau \succ A$ then $\tau \succ B$.*

The definition of support generalises the standard definition of truth in a world.

For our analysis we will also use the notion of the *truth set* of a wff A (in symbols: $|A|$). It is the set of all the worlds from $\mathcal{W}_{\mathcal{P}}$ in which A is true, where the concept of truth is understood classically (see Wiśniewski and Leszczyńska-Jasion, 2015, p. 1591).

Now we can introduce the concept of a *possibility* for a well-formed formula A. Intuitively it is a maximal state supporting A. This might be expressed as follows (see Wiśniewski and Leszczyńska-Jasion, 2015, p. 1592):

Definition 7 (Possibility) *A possibility for wff A is a state $\sigma \subseteq \mathcal{W}_{\mathcal{P}}$ such that $\sigma \succ A$ and for each $w \notin \sigma : \sigma \cup \{w\} \nsucc A$.*

I will use $\lfloor A \rfloor$ to refer to the set of all possibilities for a wff A.

In INQ we may divide all wffs into assertions and inquisitive wffs. The latter are the most interesting from our point of view, because they raise an issue to be solved. When a wff is inquisitive, the set of possibilities for that formula is comprised of at least two elements. (When a formula has only one possibility it is called an assertion.)

Let us now consider a simple example of an inquisitive formula:

$$(p \vee q) \vee \neg(p \vee q) \tag{1.1}$$

---

[5] "$\mathbf{p}$ is true in w" means "the value of $\mathbf{p}$ under w equals $\mathbf{1}$".

The set of possibilities for (1.1) is:

$$\lfloor (p \vee q) \vee \neg (p \vee q) \rfloor = \{|p|, |q|, |\neg p| \cap |\neg q|\} \tag{1.2}$$

and its union is just $\mathcal{W}_{\mathcal{P}}$.

We can also represent the possibilities for the formula (1.1) in a form of a diagram specially designed for this purpose (see Figure 1.1.). In this diagram ⑪ is the index in which both p and q are true, ⑩ is the index in which only p is true, etc. (Groenendijk and Roelofsen, 2009).

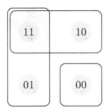

Fig. 1.1. The diagram of possibilities for the formula $(p \vee q) \vee \neg (p \vee q)$

Observe that the language $\mathcal{L}_{\mathcal{P}}$ does not include a separate syntactic category of questions. However, some wffs are regarded as *having the property of being a question*, or $\mathcal{Q}$-property for short (see Wiśniewski and Leszczyńska-Jasion, 2015, p. 1594).

Definition 8 ($\mathcal{Q}$-property) *A wff A of $\mathcal{L}_{\mathcal{P}}$ has the $\mathcal{Q}$-property iff $|A| = \mathcal{W}_{\mathcal{P}}$.*

Where $\mathcal{W}_{\mathcal{P}}$ stands for the model of $\mathcal{L}_{\mathcal{P}}$, and $|A|$ for the truth set of wff A in $\mathcal{W}_{\mathcal{P}}$. An example of a formula having the $\mathcal{Q}$-property is the formula (1.1). Hence a wff A is (i.e. has the property of being) a question just in the case when A is true in each possible world of $\mathcal{W}_{\mathcal{P}}$, the wffs having the $\mathcal{Q}$-property are just classical tautologies.

INQ adapts a "define within" approach to incorporating questions into a formal language—questions are construed as meanings of some already given well-formed formulas (see Wiśniewski and Leszczyńska-Jasion 2015, p. 1593 and Wiśniewski 2013b, p. 13).

Let Q be an initiative (i.e. a question that appears first) and Q_1 a response to the initiative. We also assume that Q and Q_1 are inquisitive formulas and that they have the $\mathcal{Q}$-property (further on I will refer to them simply as questions). $\lfloor Q \rfloor$ denotes the set of possibilities for Q. We may formally express the definition of compliance given in (Groenendijk, 2009, p. 22) as follows:

Definition 9 (Compliance)
Let $\lfloor Q \rfloor = \{|A_1|, ..., |A_n|\}$ and $\lfloor Q_1 \rfloor = \{|B_1|, ..., |B_m|\}$. Q_1 is compliant with Q (in symbols $Q_1 \propto Q$), iff

1. *For each $|B_i|$ $(1 \leq i \leq m)$ there exist $k_1, ..., k_l$ $(1 \leq k_p \leq n; 1 \leq p \leq l)$ such that $|A_{k_1}| \cup ... \cup |A_{k_l}| = \bigcup\limits_{p=1}^{l} |A_{k_p}| = |B_i|$.*

2. *For each $|A_j|$ $(1 \leq j \leq n)$ there exists $|B_k|$ $(1 \leq k \leq m)$, such that $|A_j| \subset |B_k|$.*

The intuitions behind these conditions are the following. The first one demands that every possibility in $\lfloor Q_1 \rfloor$ should be the union of some possibilities in $\lfloor Q \rfloor$—this ensures the preservation of the domain of discussion. The second condition (called a *restriction clause*) states that possibilities in $\lfloor Q \rfloor$ may only be eliminated by providing information.

As might be noticed—in the case of compliance—we cannot say anything about declarative premises involved in going from a question to a question response. The relation captured by the compliance is a subquestionhood relation. A simple example of question–question response where the response is compliant to the initiative illustrates this idea.

Example 1. Q is 'Is John coming to the party and can I come?' while Q_1 is 'Is John coming to the party'. Q may be expressed in INQ as $(p \lor \neg p) \land (q \lor \neg q)$ and Q_1 as $p \lor \neg p$. $\lfloor Q \rfloor = \{|p \land q|, |\neg p \land q|, |p \land \neg q|, |\neg p \land \neg q|\}$ and $\lfloor Q_1 \rfloor = \{|p|, |\neg p|\}$. It is the case that $Q_1 \propto Q$, because both conditions for compliance are met. For the first condition observe that $|p| = |p \land q| \cup |p \land \neg q|$ and $|\neg p| = |\neg p \land q| \cup |\neg p \land \neg q|$. For the second condition let us observe that: $|p \land q| \subset |p|$; $|p \land \neg q| \subset |p|$; $|\neg p \land q| \subset |\neg p|$; $|\neg p \land \neg q| \subset |\neg p|$.

The diagramatic representation of this example is presented in Figure 1.2.

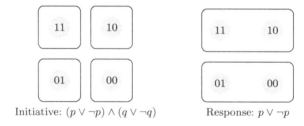

Initiative: $(p \lor \neg p) \land (q \lor \neg q)$ Response: $p \lor \neg p$

Fig. 1.2. Example of an compliant response ('Is John coming to the party?') to the question 'Is John coming to the party and can I come?' Description in the text

Compliance allows for the modelling of interesting aspects of natural language dialogues and question processing; it can be shown, however, that an erotetic implication has more an expressive power—cf. (Łupkowski, 2015). See also (Wiśniewski and Leszczyńska-Jasion, 2015) for a thorough analysis of the relations between the two paradigms: that of INQ and that of IEL.

It is worth pointing out that in the framework of inquisitive semantics it is assumed that questions and propositions are of a single ontological category.

Ginzburg et al. (2014) argue against merging questions and propositions into the same type of semantic object. The argumentation relays on two parts: the use of boolean operators in natural language and on the role of adjectives in the context of propositions and questions. See also (Wiśniewski, 2015) for a discussion of this issue.

1.2.2. Epistemic logic of questions and public announcements

Peliš and Majer (2011) employ dynamic epistemic logic of questions combined with public announcements logic (Plaza 2007, Van Ditmarsch et al. 2007) for modelling a public communication process and knowledge revisions during this process (both in the case of an individual agents' knowledge and in the case of common knowledge). This approach is based on the so-called "set-of-answers methodology" which is employed also on the grounds of IEL (see Chapter 2).

The key notion for this approach is the one of *askability*. The intuition here is to provide conditions for a situation when a question might be reasonably asked. It is important that the proposed conditions are of an epistemic character. They are:

Non-triviality.   It is not reasonable to ask a question if the answer is known.
Admissibility.   Each direct answer is considered as possible.
Context.   At least one of the direct answers must be the right one (or in other words an agent knows that one of the direct answers is the case).

Peliš and Majer (2011) present their system for S5 modal language enriched with questions, public announcement operators and group knowledge operators.
$$\varphi ::= p \mid \neg\phi \mid \phi_1 \vee \phi_2 \mid \phi_1 \wedge \phi_2 \mid \phi_1 \rightarrow \phi_2 \mid \phi_1 \leftrightarrow \phi_2 \mid K_i\phi \mid M_i\phi \mid ?_i\{\phi, ..., \phi\} \mid [\phi]\phi \mid E_G\phi \mid C_G\phi$$
The semantics of the language is a standard S5 semantics based on Kripke frames. A (Kripke) frame is a relational structure $\mathcal{F} = \langle S, R_1, ..., R_m \rangle$ with a set of states (possible worlds) S and accessibility relations $R_i \subseteq S^2$, for agents $i \in \{1, ..., m\}$. Every relation R_i is equivalence relation, i.e., reflexive, transitive, and symmetric. A (Kripke) model $\mathbf{M}$ is a pair $\langle \mathcal{F}, v \rangle$, where v is a valuation of atomic formulas.

The satisfaction relation $\models$ is defined in a standard way:

– $(\mathbf{M}, s) \models p$ iff $(\mathbf{M}, s) \in v(p)$
– $(\mathbf{M}, s) \models \neg\varphi$ iff $(\mathbf{M}, s) \not\models \varphi$
– $(\mathbf{M}, s) \models \psi_1 \wedge \psi_2$ iff $(\mathbf{M}, s) \models \psi_1$ or $(\mathbf{M}, s) \models \psi_2$
– $(\mathbf{M}, s) \models \psi_1 \vee \psi_2$ iff $(\mathbf{M}, s) \models \psi_1$ and $(\mathbf{M}, s) \models \psi_2$
– $(\mathbf{M}, s) \models \psi_1 \rightarrow \psi_2$ iff $(\mathbf{M}, s) \models \psi_1$ implies $(\mathbf{M}, s) \models \psi_2$
– $(\mathbf{M}, s) \models K_i\varphi$ iff $(\mathbf{M}, s)s \models \varphi$, for each v such that sR_is_1

The notions of satisfaction in a model or a frame and validity are defined as usual.

K_i is to be read 'the agent i knows...'. We say that

$$(\mathbf{M}, s) \models K_i\varphi \text{ iff } (\mathbf{M}, s_1) \models \varphi \text{ for each } s_1 \text{ such that } sR_is_1$$

where $\mathbf{M}$ is a Kripke model and R_i is an equivalence relation.

M_i is defined as a dual operator to K_i and means 'the agent i admits...'.

$$M_i\varphi =_{def} \neg K_i\neg\varphi$$

On top of these individual knowledge operators the following is introduced. Let us consider a group of agents $G = \{1, ..., m\}$. For this group we may define a group knowledge operator

$$E_G\varphi \leftrightarrow K_1\varphi \wedge K_2\varphi \wedge ... \wedge K_m\varphi$$

(i.e. each agent from G knows φ).

We can also introduce a stronger notion of common knowledge

$$C_G\varphi \leftrightarrow E_G\varphi \wedge E_G E_G\varphi \wedge ...$$

Intuitively, C_G means that in the group G everybody knows φ and everybody knows that everybody knows φ.

A question is the following structure:

$$?\{\alpha_1, \alpha_2, ... \alpha_n\}$$

where $\alpha_1, \alpha_2, \ldots, \alpha_n$ are formulas of the extended language. We call them direct answers of the question. Questions $?\{\alpha\wedge\beta, \alpha\wedge\neg\beta, \neg\alpha\wedge\beta, \neg\alpha\wedge\neg\beta\}$ are shortened as $?|\alpha, \beta|$ ('Is it the case that α and is it the case that β?').

Now we can express the conditions of askability in a precise manner.

Definition 10 (Askable question) *Let question Q be of the form*

$$?\{\alpha_1, \alpha_2, ... \alpha_n\}.$$

We say that Q is askable by an agent i in the state $(\mathbf{M}, s)$ (in symbols $(\mathbf{M}, s) \models Q^i$) iff

1. $(\mathbf{M}, s) \not\models K_i\alpha$ for each $\alpha \in dQ$ (non-triviality).
2. $(\mathbf{M}, s) \models M_i\alpha$ for each $\alpha \in dQ$ (admissibility).
3. $(\mathbf{M}, s) \models K_i(\alpha_1 \vee ... \vee \alpha_n)$ (context).

In the proposed framework we may also define when a question is askable for a group of agents G.

Definition 11 (Askable group question) *A question Q is an askable group question for a group of agents G in $(\mathbf{M}, s)$ (in symbols $(\mathbf{M}, s) \models G^G$) iff*

$$\forall_{i \in G}(\mathbf{M}, s) \models Q^i$$

Let us consider an example.

Example 2. We have a group of agents $G = \{a, b\}$. Their knowledge structure is represented by the following scheme:

$$
\begin{array}{ccc}
a & & b \\
\boxed{s_1} \longleftrightarrow \boxed{s_2} \longleftrightarrow \boxed{s_3} \\
\alpha & \alpha & \neg\alpha \\
\beta & \neg\beta & \neg\beta
\end{array}
$$

Agent a knows α and agent b knows $\neg\beta$. When we consider the question $?(\alpha \to \beta)$, neither of them is able to answer it by their own. This question is then askable for group $G = \{a, b\}$.

One may observe that in the presented example the situation will lead to communication between agents in order to solve the problem expressed by the question. Agents may announce their knowledge by building up the common knowledge of the group leading in consequence to obtaining the solution of the problem expressed by the question. Thus public announcements are introduced.

The public announcement operator [ ] has the following intended meaning: $[\varphi]\phi$—'after the public announcement of φ, it holds that ϕ'. The semantics of the operator is the following:

$(\mathbf{M}, s) \models [\varphi]\phi$ iff $(\mathbf{M}, s) \models \varphi$ implies $(\mathbf{M}|_\varphi, s) \models \phi$.

The model $\mathbf{M}|_\varphi$ is obtained from $\mathbf{M}$ by deleting of all states where φ is not true and by the corresponding restrictions of accessibility relations and the valuation function (for details see Peliš and Majer (2011)).

One more operator—dual to [ ]—is also introduced in (Peliš and Majer, 2011), namely $\langle \rangle$. The intended meaning of $\langle\varphi\rangle\phi$ is 'after a *truthful* announcement of φ it hold that ϕ.' We have

$$\langle\varphi\rangle\phi =_{def} \neg[\varphi]\neg\phi$$

The use of these operators allows for an analysis of the role that questions play in the multi-agent information exchange. Peliš and Majer (2011) investigate what kind of information is transmitted/revealed when an agent asks a question. For this purpose they use the notion of a *successful update*.

Definition 12 (Successful update) *A formula φ is a successful update in* $(\mathbf{M}, s)$ *iff* $(\mathbf{M}, s) \models \langle\varphi\rangle\varphi$.

Definition 13 (Successful formula) *A formula φ is a successful formula iff* $[\varphi]\varphi$ *is valid.*

It also holds that after the announcement a formula becomes common knowledge. $[\varphi]\varphi$ is valid iff $[\varphi]C_G\varphi$ is valid.

Peliš and Majer (2011) observe that in the considered formal language *questions are successful formulas.* This means that a publicly announced question becomes commonly known. As successful formulas, questions do not bring anything new if they are announced repeatedly (for proof see (Peliš and Majer, 2011)).

What is more we can show that $(\mathbf{M}, s) \models Q^i$ iff $(\mathbf{M}, s) \models \langle Q^i \rangle Q^i$, i.e. that askable questions are successful updates. As a consequence we obtain that after an agent publicly asks a question, it does not cause any change in her epistemic model. The announced question remains askable until our agent gets new information. But how does asking a question potentially influence the agents' epistemic states? This is described by the definition of an informative formula.

Definition 14 (Informative formula) *A formula φ is informative (for an agent i) with respect to Q in $(\mathbf{M}, s)$ iff $(\mathbf{M}, s) \models Q^i \wedge \langle \varphi \rangle \neg Q^i$.*

After (Peliš and Majer, 2011) let us consider an example here.

Example 3. Let us imagine a group of three card players: Ann, Bill and Catherine. Each of them has one card and nobody can see the cards of the others. Among the cards there is one Joker. Ann has the Joker but neither Bill, nor Catherine know which of the other two players has it.

Now we consider two situations which have epistemic consequences.

1. Ann announces: "I have got the Joker".
   After the public announcement of this statement an other statement holds: "Bill knows Ann has the Joker and Catherine knows Ann has the Joker."
2. The second situation is more interesting from our perspective. It is the case when Catherine says: "Who has got the Joker?". This question is askable for Catherine because

   (i)   She is not able to distinguish between the states when Bill has the Joker and when Ann has the Joker (non-triviality).
   (ii)   She considers both situations as possible (admissibility).
   (iii) The situation is such, that exactly one of the situations (Bill has the Joker or Ann has it) must be the case (context).

Observe that the question "Who has the Joker?" was also askable for Bill. However, after Catherine publicly asked this question it is no longer so. The reason for this is that Bill can infer: "I do not have the Joker and Catherine does not know who has it, therefore Ann has it". In the described situation Catherine's question is informative for Bill.

As the example shows, the presented erotetic epistemic logic allows for modelling epistemic states of sincere agents involved in a card guessing game. The task of an agent is to infer card distribution over agents on the basis of their announcements during the game. This allows for the introduction of many interesting concepts, such as askability, answerhood, partial answerhood and for building models for problems such as e.g. the Russian Cards Problem (a toy model of safe communication via an open channel)—see (Švarný et al., 2014).

Intuitions covered by the described framework will be revisited in Chapter 4 of this book. In that chapter I present a dynamic approach to erotetic search scenario execution which is inspired by the one presented by Peliš and Majer (however implemented with different logical tools). The impact of dialogic interaction on an agent's knowledge and the role of questions is modelled within dialogue logic.

1.2.3. Interrogative games and the concept of research agendas

Genot (2009a) introduces yet another interesting framework for analysing questions in interaction. The basic theories employed here are Hintikka's Interrogative Games and game theory.[6] The main objective of Genot is to apply these in the field of Belief Revision Theory presented by Olsson and Westlund (2006). The idea is to investigate an agent's questioning strategy and its modifications during an information seeking process.

From the perspective of this book the most interesting part of the presented framework are an agent's research agendas and the issue of how to preserve agenda continuity through its changes (enforced by interaction).

As a point of departure Genot (2009a) takes the notion of the *research agenda* as introduced in (Olsson and Westlund, 2006, p. 169): "a set of questions which the researcher wants to have answered.". In this approach the agenda is always relative to the so called *corpus*, which is understood as a set of sentences expressing the beliefs of an agent. Hereafter, I will refer to the corpus as K. A *question* is identified with a set of its potential answers. It is worth noticing here that only questions with finite sets of answers are considered. The disjunction of all potential answers to a given question Q is called the *presupposition* of Q.

Given the corpus K we can define K-*question* (see Olsson and Westlund 2006, p. 170 and Genot 2009a, p. 132).

Definition 15 ($K - question$) $Q = \{a_1, ..., a_n\}$ *is a* $K - question$ *iff*

1. K *entails exclusive disjunction of all elements of* Q, *and*
2. *there is no non-empty* $Q' \subset Q$ *such that* K *entails exclusive disjunction of all elements of* Q'.

Now we may introduce the K-relative research agenda—K-agenda for short (see Olsson and Westlund, 2006, p. 170).

Definition 16 A *is* K-*agenda iff* $A \subseteq Q_k$

As such K-agenda is a part of an agent's *epistemic state*

---

[6] It is also worth mentioning here the DEL_{IMI} system proposed by Hamami (2015). It combines the Interrogative Model of Inquiry with Dynamic Epistemic Logics.

$$S = \langle \underline{K}, E, \underline{A} \rangle$$

where $\underline{K}$ is the corpus, $\underline{A}$ is the research agenda and E is the so called entrenchment relation defined over $\underline{K}$.[7]

A question of the research agenda $Q \subseteq \underline{A}$ is a set of sentences (potential
answers to Q) which satisfy the following conditions (see Genot, 2009a, p. 131):

(i)   are jointly exhaustive,
(ii)   pairwise exclusive, and
(iii) non redundant given S.

An epistemic state of an agent might be modified by *expansion* (denoted by
'+') and *contraction* (denoted by '÷'). Such a change will have an effect on the
research agenda of this state.

The problem is how to preserve the continuity of agendas. Genot (2009a)
proposes to put this issue in the context of the Interrogative Game between two
agents and transform it to the task of preserving the continuity of *interrogative
strategies*. The interrogative strategy is "a list of (sequence of) questions to be
put to sources" (Genot, 2009a, p. 134). The Interrogative Game is denoted as
$\mathcal{G}(\underline{K}, Q)$ i.e. the game about Q given the background $\underline{K}$. An objective of the
game is to obtain an answer to the question Q (called the principal question).
In the game, an agent C can play:

(i)   *deductive moves* (C :!{...}—agent C states that ...);
(ii)   *interrogative moves* (C :?{...}—agent C asks whether ...).

The game $\mathcal{G}$ is presented in a form of a tree (see Genot 2009a, p. 135 and
Genot 2010, p. 246). The root of this tree is $\underline{K}$. Sign $\emptyset$ is used here to label
the leaves and it means that the complete answer to the principal question Q
is reached. Each branch of the tree represents a possible course of the game.

Genot (2009a, p. 135) defines when moves in $\mathcal{G}$ are *legal*:

– A deductive move can be added into $\mathcal{G}$ when the formula follows from $\underline{K}$
  together with other formulas on the preceding nodes (note that interrogative
  moves are not considered as formulas here).
– An interrogative move is legal if the presupposition of the question follows
  from $\underline{K}$ and the preceding nodes.

Interrogative moves are branching points of the game-tree. Each node with
an interrogative move has as many successors as there are potential answers to
the question asked.

Exemplary strategy for question $Q = \{a_1, ..., a_n\}$ is presented in Figure 1.3.

Now we may consider changes to the presented strategy which are imposed
by contraction and expansion operations.

---

[7] The core idea behind epistemic entrenchment is to express the epistemic value of the
element of a belief set. Epistemic value refers to concepts such as usefulness in inquiry or
deliberation (see Hansson, 2014).

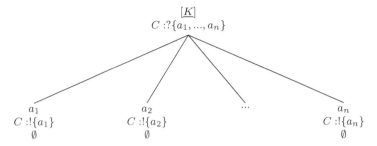

Fig. 1.3. A strategy for question $Q = \{a_1, ..., a_n\}$ (Genot, 2009a, p. 135)

Firstly, let us consider the expansion of $\underline{K}$. The intuition of expansion is that the epistemic state is expanded by additional elements in $\underline{K}$. The result is that certain questions on the state's agenda are modified so that all answers that are incompatible with the new belief state are deleted (see Olsson and Westlund, 2006, p. 172). We may say that expansion 'closes' certain questions. Figure 1.4. is an example of extending $\underline{K}$ from the figure 1.3. and its consequence on the strategy for the game $\mathcal{G}(\underline{K}, Q)$. In the presented example $\underline{K}$ is expanded with a partial answer to Q. As a result a relevant path of the tree has been removed. In the case where we would expand $\underline{K}$ with a complete answer to Q the result would be of the form presented in Figure 1.5. Observe that in this case $(\mathcal{G}(\underline{K} + a_i, Q))$ no interrogative moves are possible.

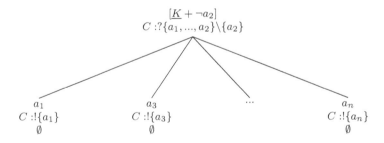

Fig. 1.4. A strategy for question $Q = \{a_1, ..., a_n\}$ after an expansion of $\underline{K}$ with a partial answer: $\neg a_2$ (Genot, 2009a, p. 136)

$$[\underline{K} + a_i]$$
$$|$$
$$C :!\{a_i\}$$
$$\emptyset$$

Fig. 1.5. A strategy for question $Q = \{a_1, ..., a_n\}$ after an expansion of $\underline{K}$ with a complete answer: a_i (Genot, 2009a, p. 136)

Now let us consider a contraction of the agent's epistemic state. After the contraction an element is removed from $\underline{K}$ and as a result the agenda has to be modified. When we remove an element α, a new question in the agenda will be introduced with α as one of its potential answers—so contraction 'opens' new questions (see Olsson and Westlund, 2006, p. 173–174). In the example presented in Figure 1.6. the strategy for the game $\mathcal{G}$ is modified after contracting $\underline{K}$ with $\neg(a_1 \wedge a_2)$.

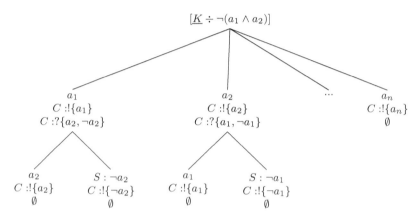

Fig. 1.6. A strategy for question $Q = \{a_1, ..., a_n\}$ after the contraction of $\underline{K}$ with $\neg(a_1 \wedge a_2)$ (Genot, 2009a, p. 136)

In what follows Genot introduces the notion of *precondition free strategy* (see Genot, 2009a, p. 137).

Definition 17 *The strategy $\sigma(\underline{K}, Q)$ (in a game $\mathcal{G}(\underline{K}, Q)$) is precondition free iff it uses only $\emptyset$-questions (i.e. $\underline{K}$-questions for $\underline{K} = \emptyset$) as interrogative moves.*

Figure 1.7. illustrates a precondition free strategy $\sigma(\underline{K}, Q = \{a_1, ..., a_n\})$.
Concerning precondition free strategies Genot (2009a, p. 137) shows that:

Proposition 1 *For any question Q, there is a precondition free interrogative strategy using only 'yes-no' questions for answering Q.*

For this proposition a *Yes-No Theorem* (see Hintikka, 1999, p. 55) is exploited.[8]

Theorem 1 (Yes-No Theorem). *For any corpus $\underline{K}$, and any $\underline{K}$-agenda $\underline{A}$, if some conclusion follows from $\underline{K}$ together with some answer to some $Q \in \underline{A}$, then it follows from $\underline{K}$ together with answers to 'yes-no' questions only.*

---

[8] For similar results on the reductibility of questions to sets of simple yes-no questions see (Wiśniewski, 1994) and (Lesniewski and Wisniewski, 2001).

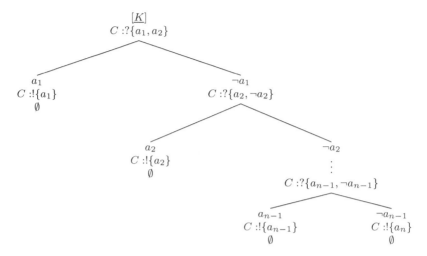

Fig. 1.7. Precondition free strategy for question $Q = \{a_1, ..., a_n\}$ (Genot, 2009a, p. 137)

And on Proposition 1 another one is built up addressing the issue of continuity of strategies (agendas) through expansion and contraction (Genot, 2009a, p.138).

Proposition 2 *Let Q be a $\underline{K}$-question and $\sigma(\underline{K}, Q)$ be a precondition free interrogative strategy in a game $\mathcal{G}(\underline{K}, Q)$. Then for any b, if Q fails to be a $\underline{K} \circ b$-question ($\circ \in \{+, \div\}$), there must be a solution to the problem of agenda continuity by simply removing from or adding to $\sigma(\underline{K}, Q)$ 'yes-no' questions.*

In (Genot, 2009b) the presented framework is extended for the case of a *revision* of $\underline{K}$ and the possibility of multiple sources of answers (i.e. multi-agent interactions). (Genot, 2010) brings more extensive and detailed discussion on the epistemic perspective on questions and techniques of updating questions after an epistemic state changes.

It is also worth mentioning the work of Genot and Jacot (2012). In this paper the issue of the informativeness of publicly asked questions is considered within the framework of Interrogative Games. Similarly to (Peliš and Majer, 2011) certain conditions concerning the situation of publicly asking questions are imposed. This is achieved by introducing a *request for information* context (rfi). If an agent C is asking a question in rfi it warrants the following attitudes of C (see Genot and Jacot, 2012, p. 3):

- C knows that at least one answer holds;
- C does not know which answer holds;
- C wants to know which answer holds.

The strategical approach to the questioning process (agent's agendas) will be addressed in the following chapter of this book (where erotetic search scenarios are introduced) and especially in Chapters 3.2 and 4 presenting erotetic search scenarios in action—i.e. modelling strategic questioning processes in the dialogical context.

1.3. Summary

In this chapter I have presented an overview of chosen frameworks for analysing the role of questions in dialogue. After Ginzburg (2011) these frameworks are presented within the domain of linguistics and logic. It is worth stressing, however, that nowadays the ideas and methods are interchanged between these broadly understood domains. Reader interested in more detailed and broader discussions might reach for (Ginzburg, 2011, 2016), (Hamami and Roelofsen, 2015) and (Wiśniewski, 2015). Ginzburg (2011, 2016) reviews logical and linguistic theories and frameworks for grasping the role of questions in dialogue while also discussing the interplay between these two approaches. Hamami and Roelofsen (2015) point out the most popular logics of questions and their results. Wiśniewski (2015) shows how questions are conceptualised in logic and formal linguistics by focusing on the most influential theories and their basic claims. What is particularly interesting Wiśniewski (2015) also discusses how a formal representation of questions reflects natural language questions.

Chapter 2

Questions and Inferential Erotetic Logic

This chapter introduces the basic concepts of Inferential Erotetic Logic. Firstly, I will present the IEL approach to formalising questions. Secondly, I will discuss the central issue of IEL—namely erotetic inferences. The chapter ends by introducing two formal languages that will be later used in this book.

IEL was developed by Andrzej Wiśniewski in the 1990s.[1] IEL is a logic which focuses on inferences whose premises and/or conclusion is a question, and which gives criteria of validity of such inferences. Thus it offers a very useful and natural framework for analyses of the questioning process.

It is worth stressing that IEL-based concepts have proven useful for many domains. These concepts have been applied in the domain of the conceptual foundations of artificial intelligence in order to address the issue of the Turing test's adequacy (see Łupkowski, 2010, 2011a; Łupkowski and Wiśniewski, 2011). IEL has also been applied in the problem solving area. Komosinski et al. (2014) use an implementation of a proof method based on IEL for a generation of abductive hypotheses. Abductive hypotheses are evaluated by multi-criteria dominance relations. We can point to IEL's application in modelling cognitive goal-directed processes (see Wiśniewski, 2003, 2001, 2012), or (Urbański and Łupkowski, 2010a).

We may also point to IEL applications in proof-theory. There are two proof-methods grounded in IEL: the method of Socratic proofs (see Wiśniewski (2004b), Leszczyńska (2004, 2007), Wiśniewski et al. (2005), Wiśniewski and Shangin (2006)) and the synthetic tableaux method (see Urbański, 2001a,b, 2002). A theorem-prover (an implementation of the method of Socratic proofs for Classical Propositional Logic is described in Wiśniewski (2004b)) implemented in Prolog by Albrecht Heefer and available at: http://logica.ugent.be/albrecht/socratic.html. A theorem-prover (an implementation of the method of Socratic proofs for 15 basic propositional modal logics described in Leszczyńska (2004, 2007)) written in Prolog by Albrecht Heefer and Dorota Leszczyńska-Jasion, is available at: http://logica.ugent.be/albrecht/socratic-modal.htm.

---

[1] For an extensive IEL presentation see (Wiśniewski, 1995) and (Wiśniewski, 2013b).

It is worth noticing that IEL is also used as a theoretical background in the context of empirical research. Moradlou and Ginzburg (2014) present a corpus study aimed at characterising the learning process by means of which children learn to understand questions. The authors assume that for some stages of this process children are attuned to a very simple *erotetic logic*. Urbański et al. (2014) present research on correlations between the level of fluid intelligence and fluencies in two kinds of deductions: simple (syllogistic reasoning) and dificult ones (erotetic reasoning). The tool used to investigate erotetic reasoning is the *Erotetic Reasoning Test* which expoits IEL concepts (such as e-implication). For more discussion on this subject see also (Urbański et al., 2016a). Łupkowski and Wietrzycka (2015) (see also Łupkowski et al. 2015) present a game with a purpose, where IEL concepts are used as a normative yardstick for game design.

2.1. Questions

When it comes to formalising questions, IEL employs the so called set-of-answers methodology (see Harrah 2002; Wiśniewski 2013b, p. 16–18; Peliš 2010; Peliš 2016). This approach is rooted in the Hamblin's postulate (Hamblin, 1958, p. 162) stating that

(2)   Knowing what counts as an answer is equivalent to knowing the question.

Wiśniewski (2013b) describes this approach as a semi-reductionistic one. After Wiśniewski (2013b) it is worth introducing the distinction between:

– natural language-questions (NLQ), i.e. questions formulated in a given natural language; and
– erotetic formulas (hereafter e-formulas), which are formulas of a certain formal language, that represent NLQs.

How can we understand the nature of this 'representation'? The key to this notion lies in the answers. As we read in (Wiśniewski, 2013b, p. 15)

An e-formula Q represents a NLQ Q^* construed in such a way that possible answers to Q^* having the desired semantic and/or pragmatic properties are represented/formalized by direct answers to Q.

Direct answers are understood as possible just-sufficient answers. Such a direct answer "gives exactly what the question calls for" (Harrah, 2002, p. 1).

The distinction between NLQs and e-formulas allows for a more precise formulation of the postulate 2 (see Wiśniewski, 2013b, p. 16):

(3) a. Knowing a NLQ is equivalent to knowing the e-formula that represents it.
     b. Knowing the e-formula is equivalent to knowing what counts as a direct answer to it.

Placing the focus on answers opens the possibility of a syntactic approach to e-formulas. When we enrich a language with symbols, like the question mark ? and brackets {, } we can express e-formulas in the following form:

$$?\{A_1, A_2, \ldots, A_n\} \tag{2.1}$$

where $n > 1$ and $A_1, A_2, \ldots, A_n$ are nonequiform, that is, pairwise syntactically distinct, declarative well-formed formulas (d-wffs) of a given formal language. After (Wiśniewski, 2013b, p. 18) I will refer to e-formulas of formal languages as questions of these languages. If $?\{A_1, A_2, \ldots, A_n\}$ is a question, then each of the d-wffs $A_1, A_2, \ldots, A_n$ is a direct answer to the question.

A question $?\{A_1, A_2, \ldots, A_n\}$ can be read, 'Is it the case that A_1, or is it the case that A_2, ..., or is it the case that A_n?'.

In general IEL accepts the following two assumptions with questions (see Wiśniewski, 2013b, p. 57):

- $(\mathbf{sc}_1)$ direct answers are sentences;
- $(\mathbf{sc}_2)$ each question has at least two direct answers.

IEL introduces a series of semantic concepts about questions. Semantics for questions are provided by the means of the so called Minimal Erotetic Semantics (MiES)—for more details see (Wiśniewski, 2013b, Chapter 4). I will present and discuss several of these concepts later in this chapter in the context of the already introduced formal languages.

2.2. Erotetic inferences

In this section I will provide an informal description and intuitions behind a certain kind of erotetic inferences considered within IEL. Formal definitions will be given in Section 2.3.

2.2.1. Erotetic implication

In IEL erotetic inferences of two kinds are analysed:

- *Erotetic inferences of the first kind*, where a set of premises consists of declarative sentence(s) only, and an agent passes from it to a question—grasped under the notion of *question evocation* (see Wiśniewski, 2013b, Chapter 6).
- *Erotetic inferences of the second kind*, where a set of premises consists of a question and possibly some declarative sentence(s) and an agent passes from it to another question—grasped under the notion of *erotetic implication*.

In this book I will be interested only in the erotetic inferences of the second kind. E-implication is a semantic relation between a question, Q, a (possibly empty) set of declarative well-formed formulas, X, and a question, Q_1. It is an ordered triple $\langle Q, X, Q_1 \rangle$, where Q is called an *interrogative premise* or simply *initial question*, the elements of X are *declarative premises* and the question Q_1 is the *conclusion* or the *implied question*—see (Wiśniewski, 2013b, p. 51–52).

The intuition behind e-implication might be expressed as follows. Let us imagine an agent who is trying to solve a certain (possibly) complex problem. The problem is expressed by her initial question (Q). We assume that the agent does not have resources to answer the initial question on her own. Thus the initial question has to be processed/decomposed. This decomposition is aimed at replacing the initial question with a simpler auxiliary question—Q_1. The auxiliary question obtained as a result of the decomposition process should have certain characteristics. First of all, it should stay on the main topic. In other words, no random questions should appear here (as in the questioning agenda intuitions described in Chapter 1.2.3). However, the main characteristic that we are aiming at here is that the answer provided to the auxiliary question should be at least a partial answer to the initial question (i.e. it should narrow down the set of direct answers to the initial question, see Wiśniewski 2013b, p. 43). It should bring our agent closer to solving the initial problem. Summing up, we can perceive the discussed process of replacing one question with another (simpler) one as a well-motivated step from the problem-solving perspective. In our case, it is e-implication, which provides a normative yardstick for the question decomposition we are querying.

Let us consider (after Urbański and Łupkowski, 2010b, p. 68) a simple example of e-implication in action.

Example 4. Let us imagine that our initial problem is expressed by the following question:

(Q) Who stole the tarts?

Suppose also that we have managed to establish the following evidence:

(E_1) One of the courtiers of the Queen of Hearts attending the afternoon tea-party stole the tarts.

We will use this piece of knowledge for the purpose of the initial question decomposition. One may notice that the evidence E_1 allows for the formulation of an auxiliary question that will be easier to answer than the initial one. E_1 states that to answer Q one would not have to search for the answer concerning all the possibilities, but rather restrict one's interests to the courtiers of the Queen of Hearts attending the afternoon tea-party. Thus our initial question together with the evidence E_1 implies the question:

(Q_1) Which of the Queen of Hearts' courtiers attended the afternoon tea-party?

Q_1 restricts the search space to the Queen of Hearts' courtiers attending the party (i.e. it asks for more specific information than the initial question). What is more, providing the answer to Q_1 will certainly be helpful in obtaining the answer to the initial question Q. We may imagine that the answer to the Q_1 will be some sort of list of the courtiers. This will be our new piece of evidence that will allow for the formulation of yet another auxiliary question(-s) concerning courtiers from the list.

IEL provides the following conditions of validity for the described situation (see Wiśniewski, 2013b, p. 52–53) making our intuitions more precise:

(4)   a. If the initial question has at least one true answer (with respect to the underlying semantics) and all the declarative premises are true, then the question which is the conclusion must have at least one true answer.

   b. For each direct answer B to the question which is the conclusion there exists a non-empty proper subset Y of the set of direct answers to the initial question such that the following condition holds:
   (♣) if B is true and all the declarative premises are true, then at least one direct answer $A \in Y$ to the initial question must be true.

2.2.2. Erotetic Search Scenarios

When we think about e-implication used for decomposing questions as described above it is easy to imagine that it might be repetitively applied while solving a particular complex problem. The intuition behind such a process is perfectly grasped under:

EDP   (*Erotetic Decomposition Principle*) Transform a principal question into auxiliary questions in such a way that: (a) consecutive auxiliary questions are dependent upon the previous questions and, possibly, answers to previous auxiliary questions and, (b) once auxiliary questions are resolved, the principal question is resolved as well (Wiśniewski, 2013b, p. 103).

This leads us to the notion of an *erotetic search scenario* (e-scenario in short). As the name suggests it is a scenario for solving a problem expressed in the form of a question. The pragmatic intuition behind the e-scenario is that it

> (...) provides information about possible ways of solving the problem expressed by its principal question: it shows what additional data should be collected if needed and when they should be collected. What is important, an e-scenario provides the appropriate instruction for every possible and just-sufficient, i.e. direct answer to a query: there are no "dead ends". (Wiśniewski, 2013a, p. 110).

Figure 2.1. presents the e-scenario structure in a schematic form. Figure 2.2. presents a natural language example of a questioning plan which has the structure of an e-scenario.

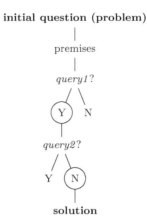

Fig. 2.1. An exemplary structure of an e-scenario. Description in the text

The e-scenario has a tree-like structure with the main question as the root and direct answers to it as leaves. Other questions are auxiliary. In our examples (Figures 2.1. and 2.2.) the initial question—the main problem for these e-scenarios are typeset with a bold font, the auxiliary questions are in italics. The premises used are displayed between these questions.

An auxiliary question has another question as the immediate successor or it has all the direct answers to it as the immediate successors. In the latter case, the immediate successors represent the possible ways in which the relevant request for information can be satisfied, and the structure of the e-scenario shows what further information requests (if any) are to be satisfied in order to arrive at an answer to the main question. If an auxiliary question is a 'branching point' of an e-scenario, it is called a *query* of the e-scenario. Among auxiliary questions, only queries are to be asked; the remaining auxiliary questions serve as erotetic premises only.

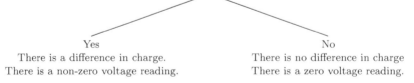

Fig. 2.2. Example of a questioning plan with an e-scenario structure. The example is based on a tutor-student dialogue from The Basic Electricity and Electronics Corpus (Rosé et al., 1999), file BEE(F), stud37

An e-scenario might be viewed as providing a search plan for an answer to the initial question. This plan is relative to the premises a questioner has, and leads through auxiliary questions (and the answers to them) to the answer to the initial question. Each path of an e-scenario leading from the root to one of the leaves represents one of the ways in which the process of solving the initial problem might go. This allows us to consider issues referred to as *distributed internal question processing* (see Wiśniewski, 2013b, p. 105).

The key feature of e-scenarios is that auxiliary questions appear in them on the condition that they are e-implied. Thus we may use e-scenarios to provide some insights into questioning strategies used in dialogues. This approach is efficient for contexts where a questioner wants to obtain an answer to the initial question, which should not be asked directly (as e.g. in the Turing test situation, where asking a direct question 'Are you a human or a machine?' would be fruitless as a satisfactory way of obtaining a solution to the problem of agent identification).[2] To obtain an answer to the initial question, the questioner usually asks a series of auxiliary questions in these situations. Answers to these questions build up to be an answer to the initial one. It is easy to imagine a context such as this in real life situations, as for example while teaching, when we want to check if our student really understands a given problem.

2.3. Formal languages to use

Now I may introduce formal languages that will be used further on in this book. I will also provide formal definitions for concepts presented at the intuitive level earlier in this chapter.

For the purposes of this book I will only need propositional languages with questions. The reason for this is that the expressive power of the introduced languages is just-sufficient for the analysis presented. It is worth stressing that MiES enables for enriching any formal language with questions, provided that this language allows for partitioning declarative formulas into true and untrue ones (cf. Wiśniewski 1996, 2001, 2013b; Peliš 2011, 2016).

In what follows I will use Q, Q^*, Q_1, ... as metalinguistic variables for questions and A, B, C, D, possibly with subscripts, as metalinguistic variables for declarative well-formed formulas, $X, Y, \ldots$ represent sets of d-wffs. I will use dQ for the set of direct answers to a question Q.

---

[2] See Łupkowski (2011a), Urbański and Łupkowski (2010b).

2.3.1. Language $\mathcal{L}^?_{cpl}$

Following Wiśniewski (2013b, chapter 2) I present language $\mathcal{L}^?_{cpl}$. Let us start with $\mathcal{L}_{cpl}$ which is the language of Classical Propositional Logic (CPL, for short) Language $\mathcal{L}_{cpl}$ contains the following primitive connectives: $\neg$ (negation), $\rightarrow$ (implication), $\vee$ (disjunction), $\wedge$ (conjunction), $\leftrightarrow$ (equivalence). The concept of a *well-formed formula* (wff for short) is defined in a traditional manner. We assume that $\vee$ and $\wedge$ bind more strongly than $\rightarrow$ and $\leftrightarrow$. We use p, q, r, s, p_1, ... for propositional variables. CPL-valuation (v) is understood in a standard way.

At this point, I introduce another object-level language—$\mathcal{L}^?_{cpl}$. The vocabulary of the new language is the vocabulary of $\mathcal{L}_{cpl}$ extended with the following signs: ?, {, }, and the comma. As described in Section 2.1 this allows us to represent the erotetic formulas (e-formulas) of the language. Consequently we say that $\mathcal{L}^?_{cpl}$ has two categories of well-formed expressions: declarative well-formed formulas (hereafter d-wffs) and erotetic well-formed formulas (i.e. questions, hereafter e-wffs). The categories of d-wffs and e-wffs are disjoint. D-wffs of $\mathcal{L}^?_{cpl}$ are simply well-formed formulas of $\mathcal{L}_{cpl}$, and e-wffs of $\mathcal{L}^?_{cpl}$ are expressions of the form:

$$?\{A_1, \ldots, A_n\} \qquad (2.2)$$

where $n > 1$ and $A_1, \ldots, A_n$ are nonequiform (i.e. pairwise syntactically distinct) d-wffs of $\mathcal{L}^?_{cpl}$ (i.e. CPL-wffs). If $?\{A_1, \ldots, A_n\}$ is a question, then each of the d-wffs $A_1, \ldots, A_n$ is called a *direct answer* to the question. As we can see, each question of $\mathcal{L}^?_{cpl}$ has a finite set of direct answers and each question has at least two direct answers.[3]

Any question of the form (2.2) may be read:

Is it the case that A_1, or ..., or is it the case that A_n?

In what follows, for the sake of simplicity I will adopt some notational conventions. A simple yes-no question (i.e. a question whose set of direct answers consists of a sentence and its classical negation) of the form

$$?\{A, \neg A\} \qquad (2.3)$$

are simply presented as:

$$?A \qquad (2.4)$$

Questions of the form (2.4) can be read:

Is it the case that A?

---

[3] It is worth mentioning that in IEL also other types of questions (including the ones with infinite sets of possible answers) are considered—see (Wiśniewski, 1995, Chapter 3).

Later on I will also make use of questions of the following form (2.5), called *binary conjunctive questions*[4]:

$$?\{A \wedge B, A \wedge \neg B, \neg A \wedge B, \neg A \wedge \neg B\} \tag{2.5}$$

which will be abbreviated as

$$? \pm |A, B| \tag{2.6}$$

Before I provide formal definitions of e-implication and e-scenario I will introduce the necessary concepts of MiES. The basic semantic notion to be used here is that of a *partition* (see Wiśniewski, 2013b, p. 25–30).

Definition 18 (Partition of the set of d-wffs) *Let* $\mathcal{D}_{\mathcal{L}^?_{cpl}}$ *designate the set of d-wffs of* $\mathcal{L}^?_{cpl}$. *A partition of* $\mathcal{D}_{\mathcal{L}^?_{cpl}}$ *is an ordered pair:*

$$\mathsf{P} = \langle \mathsf{T}_\mathsf{P}, \mathsf{U}_\mathsf{P} \rangle$$

where $\mathsf{T}_\mathsf{P} \cap \mathsf{U}_\mathsf{P} = \emptyset$ *and* $\mathsf{T}_\mathsf{P} \cup \mathsf{U}_\mathsf{P} = \mathcal{D}_{\mathcal{L}^?_{cpl}}$.

Intuitively, T_P consists of all d-wffs which are true in P, and U_P is made up of all the d-wffs which are untrue in P (see Wiśniewski 2013b, p. 25). If for a certain partition P and a d-wff A, $A \in \mathsf{T}_\mathsf{P}$, then we say that A is *true in partition* P, otherwise, A is *untrue in* P.

Definition 19 (Partition of the language $\mathcal{L}^?_{cpl}$) *By a partition of the language* $\mathcal{L}^?_{cpl}$ *we mean a partition of* $\mathcal{D}_{\mathcal{L}^?_{cpl}}$.

The concept of the partition is very general, thus Wiśniewski (2013b, p. 26,30) introduces the class of admissible partitions being a non empty subclass of all partitions of the language. This step allows for defining useful semantic concepts.

Definition 20 (Admissible partitions of $\mathcal{L}^?_{cpl}$) *A partition* $\mathsf{P} = \langle \mathsf{T}_\mathsf{P}, \mathsf{U}_\mathsf{P} \rangle$ *of* $\mathcal{L}^?_{cpl}$ *is admissible iff for some CPL-valuation* v:

(i) $\mathsf{T}_\mathsf{P} = \{A \in \mathcal{D}_{\mathcal{L}^?_{cpl}} : v(A) = \mathbf{1}\}$, *and*
(ii) $\mathsf{U}_\mathsf{P} = \{B \in \mathcal{D}_{\mathcal{L}^?_{cpl}} : v(B) = \mathbf{0}\}$.

The set of truths of an admissible partition of $\mathcal{L}^?_{cpl}$ equals the set of d-wffs which are true under the corresponding CPL-valuation.

Partitioning of the language concerns only declarative formulas. A question is neither in T_P nor in U_P, for any partition P—MiES does not presuppose that questions are true or false (Wiśniewski, 2013b, p. 26). As a counterpart of truth for declarative formulas, for questions we introduce the notion of *soundness* (see Wiśniewski, 2013b, p. 37).

---

[4] For a general definition of the conjunctive questions (see Urbański, 2001a, p. 76).

Definition 21 (Soundness) *A question Q is sound in a partition* P *iff* $\mathsf{d}Q \cap \mathsf{T_P} \neq \emptyset$.

A question is sound (in a partition) iff at least one direct answer to this question is true in the partition.

If a question is sound in each admissible partition P of a language it is called a *safe* question (see Wiśniewski, 2013b, p. 38).

Definition 22 (Safety) *A question Q is safe iff* $\mathsf{d}Q \cap \mathsf{T_P} \neq \emptyset$ *for each admissible partition* P.

2.3.1.1. E-implication revisited

Before I introduce the definition of erotetic implication the definition of multiple-conclusion entailment (mc-entailment)[5] is needed (Wiśniewski, 2013b, p. 33).

Definition 23 (Multiple-conclusion entailment) *Let X and Y be sets of d-wffs of language $\mathcal{L}^?_{cpl}$. We say that X mc-entails Y in $\mathcal{L}^?_{cpl}$ (in symbols $\models_{\mathcal{L}^?_{cpl}} Y$) iff there is no admissible partition $\mathsf{P} = \langle \mathsf{T_P}, \mathsf{U_P} \rangle$ of $\mathcal{L}^?_{cpl}$ such that $X \subseteq \mathsf{T_P}$ and $Y \subseteq \mathsf{U_P}$*

The intuition behind mc-entailment is that it holds between the sets of d-wffs X and Y iff the truth of all d-wffs in X warrants the presence of at least one true d-wff in Y. One may notice that this allows us to formally express the idea behind the condition (♣) expressed in page 27.

Now we may introduce the definition of erotetic implication (see Wiśniewski, 2013b, p. 68).

Definition 24 (Erotetic implication) *A question Q implies a question Q_1 on the basis of a set of d-wffs X (in symbols, $\mathsf{Im}(Q, X, Q_1)$) iff:*

(1)   for each $A \in \mathsf{d}Q$: $X \cup \{A\} \models_{\mathcal{L}^?_{cpl}} \mathsf{d}Q_1$, and

(2)   for each $B \in \mathsf{d}Q_1$ there exists a non-empty proper subset Y of $\mathsf{d}Q$ such that $X \cup \{B\} \models_{\mathcal{L}^?_{cpl}} Y$

The first clause of the above definition warrants the *transmission of soundness* (of the implying question Q) *and truth* (of the declarative premises in X) *into soundness* (of the implied question Q_1). The second clause expresses the property of "open-minded cognitive usefulness" of e-implication, that is, the fact that each answer to the implied question Q_1 narrows down the set of direct answers to the implying question Q.

If a set X of declarative formulas is empty, an e-implication of this sort is called a *pure* e-implication (see Wiśniewski, 2013b, p. 76).

---

[5] See (Shoesmith and Smiley, 1978).

Definition 25 (Pure erotetic implication) *A question Q implies a question Q_1 (in symbols, $\mathsf{Im}(Q, Q_1)$) iff:*

(1)   for each $A \in \mathsf{d}Q$: $A \Vvdash_{\mathcal{L}^?_{cpl}} \mathsf{d}Q_1$, and

(2)   for each $B \in \mathsf{d}Q_1$ there exists a non-empty proper subset Y of $\mathsf{d}Q$ such that $B \Vvdash_{\mathcal{L}^?_{cpl}} Y$

Let us now consider several examples of e-implication, starting with a pure one.

$$\mathsf{Im}(?\{A, B \vee C\}, ?\{A, B, C\}) \qquad (2.7)$$

In (2.7) Q is $?\{A, B \vee C\}$ and Q_1 is $?\{A, B, C\}$. The first condition for a pure e-implication is met. The same applies to the second condition. One may observe that the proper subset Y of the set of direct answers to the question Q is the following: (i) for the direct answer A to question Q_1 it is $\{A\}$, (ii) when it comes to the answer B it is $\{B \vee C\}$, and (iii) for the answer C it is also $\{B \vee C\}$.

$$\mathsf{Im}(?A, A \leftrightarrow B, ?B) \qquad (2.8)$$

In (2.8) we may notice that also two conditions of e-implication are met. $?A$ is a simple yes-no question, thus the set of direct answers to this question is $\{A, \neg A\}$. The set of direct answers to the implied question $?B$ is $\{B, \neg B\}$. For each direct answer to $?A$ if it is true and the premise is true, then at leas one direct answer to $?B$ is true (it is B for A and $\neg B$ for $\neg A$). As for the second condition of the e-implication it is also met. The required proper subset Y of the set of direct answers to the implying question $?A$ is the following: (i) for the direct answer B to the question $?B$ it is $\{A\}$, and (ii) for the direct answer $\neg B$ to the question $?B$ it is $\{\neg A\}$.

After Wiśniewski (2013b, p. 78–80) we may also introduce several other examples.

$$\mathsf{Im}(?\neg A, ?A) \qquad (2.9)$$

$$\mathsf{Im}(?A, ?\neg A) \qquad (2.10)$$

$$\mathsf{Im}(?\{A, B\}, ?\{B, A\}) \qquad (2.11)$$

$$\mathsf{Im}(? \pm |A, B|, ?A) \qquad (2.12)$$

$$\mathsf{Im}(?B, ? \pm |A, B|) \qquad (2.13)$$

$$\mathsf{Im}(? \pm |A, B|, ?(A \otimes B)) \qquad (2.14)$$

where $\otimes$ is any of the connectives: $\wedge$, $\vee$, $\rightarrow$, $\leftrightarrow$.

$$\mathsf{Im}(?(A \otimes B), ? \pm |A, B|) \tag{2.15}$$

where $\otimes$ is any of the connectives: $\wedge, \vee, \rightarrow, \leftrightarrow$.

$$\mathsf{Im}(?A, B \rightarrow A, ?\{A, \neg A, B\}) \tag{2.16}$$

$$\mathsf{Im}(?A, A \rightarrow B, ?\{A, \neg A, \neg B\}) \tag{2.17}$$

$$\mathsf{Im}(?\{A, B, C\}, D \rightarrow B, \neg D \rightarrow B \vee C, ?D) \tag{2.18}$$

$$\mathsf{Im}(?\{B, C\}, \neg D \rightarrow B \vee C, \neg D, B \leftrightarrow E, ?E) \tag{2.19}$$

When examples (2.16) and (2.17) are considered we should mention some important characteristics of e-implication. It is the case that the relation of e-implication is not transitive. As a consequence there are cases where more than one erotetic reasoning step is needed to access one question from the other one using e-implication. For example, although $?A$ does not e-imply $?B$, $?B$ is accessible from $?A$ on the basis of $B \rightarrow A$ in *two steps*:

$$\mathsf{Im}(?A, B \rightarrow A, ?\{A, \neg A, B\}) \tag{2.20a}$$
$$\mathsf{Im}(?\{A, \neg A, B\}, ?B) \tag{2.20b}$$

We will come across such multi-step transitions in the following chapters of this book (especially within the context of e-scenarios). For more discussion about this e-implication feature see (Wiśniewski, 2013b, Chapter 7).

2.3.1.2. E-scenarios revisited

We are now ready to introduce the definition of an e-scenario. Here—after Wiśniewski (2013b)—I will present the e-scenario as a family of interconnected sequences of the so-called erotetic derivations.[6] It is worth mentioning that e-scenarios can also be viewed as labelled trees (see Leszczyńska-Jasion, 2013).

Erotetic derivation is defined as follows (Wiśniewski, 2013b, p. 110–111):

Definition 26 (Erotetic derivation) *A finite sequence* $\mathbf{s} = \mathbf{s}_1, \dots, \mathbf{s}_n$ *of wffs is an erotetic derivation (e-derivation for short) of a direct answer* A *to question* Q *on the basis of a set of d-wffs* X *iff* $\mathbf{s}_1 = Q$, $\mathbf{s}_n = A$, *and the following conditions hold:*

(1)   for each question $\mathbf{s}_k$ *of* $\mathbf{s}$ *such that* $k > 1$:

---

[6] See also (Wiśniewski, 2001) and (Wiśniewski, 2003) where the idea of e-scenarios has been presented for the first time.

(a) $\mathbf{ds}_k \neq \mathbf{d}Q$,
(b) $\mathbf{s}_k$ *is implied by a certain question* $\mathbf{s}_j$ *which precedes* $\mathbf{s}_k$ *in* $\mathbf{s}$ *on the basis of the empty set, or on the basis of a non-empty set of d-wffs such that each element of this set precedes* $\mathbf{s}_k$ *in* $\mathbf{s}$, *and*
(c) $\mathbf{s}_{k+1}$ *is either a direct answer to* $\mathbf{s}_k$ *or a question;*

(2)   *for each d-wff* $\mathbf{s}_i$ *of* $\mathbf{s}$:

(a) $\mathbf{s}_i \in X$, *or*
(b) $\mathbf{s}_i$ *is a direct answer to* $\mathbf{s}_{i-1}$, *where* $\mathbf{s}_{i-1} \neq Q$, *or*
(c) $\mathbf{s}_i$ *is entailed by a certain non-empty set of d-wffs such that each element of this set precedes* $\mathbf{s}_i$ *in* $\mathbf{s}$;

An e-derivation is *goal-directed*: it leads from an initial question Q to a direct answer to this question.Clause (1a) of the above definition requires that an auxiliary question (i.e. a question of an e-derivation different from Q) appearing in an e-derivation should have different direct answers than the initial question Q. Clause (1b) amounts to the requirement that each question of the e-derivation which is different from the initial question Q must be e-implied by some earlier item(s) of the e-derivation. Clause (1c) requires that an immediate successor of an auxiliary question in the e-derivation must be a direct answer to that question or a further auxiliary question. Clause (2) enumerates reasons for which a d-wff may enter an e-derivation. Such a d-wff may be: (2a) an element of a set of d-wffs X; (2b) a direct answer to an auxiliary question; (2c) a consequence of earlier d-wffs.

Definition 27 (Erotetic search scenario) *A finite family* Σ *of sequences of wffs is an erotetic search scenario (e-scenario for short) for a question* Q *relative to a set of d-wffs* X *iff each element of* Σ *is an e-derivation of a direct answer to* Q *on the basis of* X *and the following conditions hold:*

(1) $\mathbf{d}Q \cap X = \emptyset$;
(2) Σ *contains at least two elements;*
(3)   *for each element* $\mathbf{s} = \mathbf{s}_1, \ldots, \mathbf{s}_n$ *of* Σ, *for each index* k, *where* $1 \leq k < n$:

(a)   *if* $\mathbf{s}_k$ *is a question and* $\mathbf{s}_{k+1}$ *is a direct answer to* $\mathbf{s}_k$, *then for each direct answer* B *to* $\mathbf{s}_k$: *the family* Σ *contains a certain e-derivation* $\mathbf{s}^* = \mathbf{s}_1^*, \mathbf{s}_2^*, \ldots, \mathbf{s}_m^*$ *such that* $\mathbf{s}_j = \mathbf{s}_j^*$ *for* $j = 1, \ldots, k$, *and* $\mathbf{s}_{k+1}^* = B$;
(b)   *if* $\mathbf{s}_k$ *is a d-wff, or* $\mathbf{s}_k$ *is a question and* $\mathbf{s}_{k+1}$ *is not a direct answer to* $\mathbf{s}_k$, *then for each e-derivation* $\mathbf{s}^* = \mathbf{s}_1^*, \mathbf{s}_2^*, \ldots, \mathbf{s}_m^*$ *in* Σ *such that* $\mathbf{s}_j = \mathbf{s}_j^*$ *for* $j = 1, \ldots, k$ *we have* $\mathbf{s}_{k+1} = \mathbf{s}_{k+1}^*$.

E-derivations being elements of an e-scenario will be called *paths* of this e-scenario.

For my purposes notions of *query of an e-derivation* (Wiśniewski, 2013b, p. 112) and *query of an e-scenario* (Wiśniewski, 2013b, p. 113) will also be needed.

Definition 28 (Query of an e-derivation) *A term s_k (where $1 < k < n$) of an e-derivation $s = s_1, \ldots s_n$ is a query of s if s_k is a question and s_{k+1} is a direct answer to s_k.*

Definition 29 (Query of an e-scenario) *A query of an e-scenario is a query of a path of the e-scenario.*

It is worth noticing that (see Wiśniewski, 2013b, p. 114–115):

1. each e-scenario has the first query shared by all the paths of the scenario;
2. each query of an e-scenario is a question with a finite number of direct answers;
3. each path of an e-scenario involves at least one query.

As an illustration of the above concepts, let us consider a simple example— see Figure 2.3. The initial question of our exemplary e-scenario is $?p$. Only one declarative premise is employed here, namely $p \leftrightarrow q$. This e-scenario contains two paths (i.e. two e-derivations):

$$?p, p \leftrightarrow q, ?q, q, p \tag{2.21a}$$
$$?p, p \leftrightarrow q, ?q, \neg q, \neg p \tag{2.21b}$$

The e-scenario has only one query, i.e. $?q$. The query is e-implied by the initial question and the declarative premise (see e-implication scheme (2.8)).

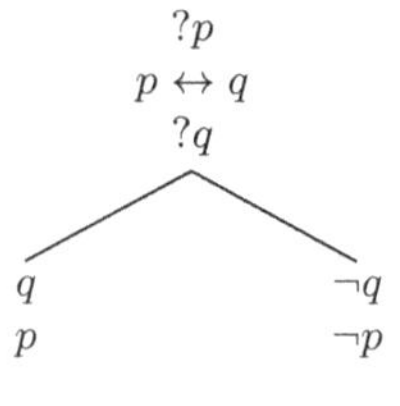

Fig. 2.3. Example of an e-scenario for the question $?p$ relative to the premise $p \leftrightarrow q$

Figures 2.4. and 2.5. present how e-scenario might get more complex by extending sets of declarative premises.

We may also consider an e-scenario with an empty set of declarative premises. Such a scenario is called a *pure one* (Wiśniewski, 2013b, p. 128).

Definition 30 (Pure e-scenario) *A pure e-scenario is an e-scenario which does not involve any initial declarative premise.*

An exemplary scenario of this kind is presented in Figure 2.6.

As might be noticed in the presented examples (Figures 2.3., 2.4. and 2.5., 2.6.) it is the case that each answer to the initial question of the e-scenario is (a label of) a certain leaf. E-scenarios with this type of feature are called *complete* (Wiśniewski, 2013b, p. 126).

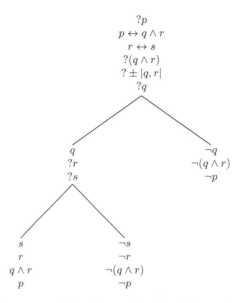

Fig. 2.4. Example of an e-scenario for the question $?p$ relative to the set of premises $\{p \leftrightarrow q \wedge r, r \leftrightarrow s\}$

Definition 31 (Complete e-scenario) *An e-scenario Σ for Q relative to X is complete if each direct answer to Q is the last term of a path of Σ; otherwise Σ is incomplete.*

Yet another structural feature shared by all the presented e-scenarios is the placement of declarative initial premises. This takes us to the notion of the *canonical form* of an e-scenario (cf. Wiśniewski, 2013b, p. 122).

Definition 32 (E-scenario in the canonical form) *An e-scenario Σ for Q relative to X is in the* cannonical form *when all the initial premises occur in Σ before the first query.*

Yet another type of e-scenario will be useful for the needs of this book. These will be the so called *information-picking* e-scenarios. First we will introduce an auxiliary notion of a informative relative question (Wiśniewski, 2013b, p. 134).

Definition 33 (Informative relative question) *We say that a question Q is informative relative to a set of d-wffs Z iff no direct answer to Q is entailed by Z.*

We also introduce the set $dec_{\mathbf{s}}^{<}(\mathbf{s}_k)$ which consists of terms of a path of an e-scenario $\mathbf{s}$ that occur in $\mathbf{s}$ before a given element of this path—$\mathbf{s}_k$. Let $\mathbf{s} = \mathbf{s}_1, ..., \mathbf{s}_n$ be a path of an e-scenario Σ. We define:

$$dec_{\mathbf{s}}^{<}(\mathbf{s}_k) = \{\mathbf{s}_j : j < k \text{ and } \mathbf{s}_j \text{ is a d-wff }\}$$

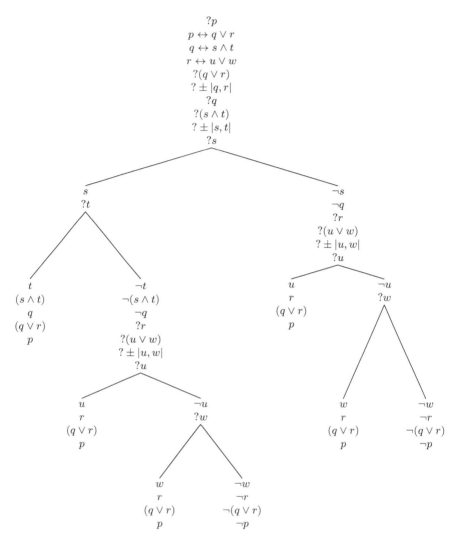

Fig. 2.5. Example of an e-scenario for the question $?p$ relative to the set of premises $\{p \leftrightarrow q \vee r, q \leftrightarrow s \wedge t, r \leftrightarrow u \vee w\}$

Now we are ready to introduce a definition of information-picking e-scenario (Wiśniewski, 2013b, p. 134).

Definition 34 (Information-picking e-scenario) *An e-scenario Σ for Q relative to X is information-picking iff:*

1. Q is informative relative to the set of initial premises of Σ, and
2. for each path $\mathbf{s} = \mathbf{s}_1, ..., \mathbf{s}_n$ of Σ:

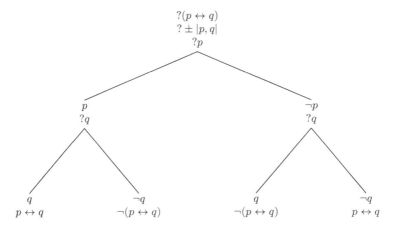

Fig. 2.6. An exemplary pure e-scenario for the question $?(p \leftrightarrow q)$

    a. if $\mathbf{s}_k$ is a query of $\mathbf{s}$, then $\mathbf{s}_k$ is informative relative to $dec^{<}_{\mathbf{s}}(\mathbf{s}_k)$, and
          $\mathbf{s}_{k+1} \notin \mathsf{d}Q$,
    b. $dec^{<}_{\mathbf{s}}(\mathbf{s}_n)$ entails $\mathbf{s}_n$.

    The main intuition behind this class of e-scenarios is that the initial question cannot be resolved by means of the initial premises only. Additional pieces of information have to be collected by means of answering queries.

    Taking into account their applications, one of the most important properties of e-scenarios is expressed by the *Golden Path Theorem*. I present here a somewhat simplified version of the theorem (for the original see Wiśniewski 2003, p. 411 or Wiśniewski 2013b, p. 126):

Theorem 2 (Golden Path Theorem). *Let Σ be an e-scenario for a question Q relative to a set of d-wffs X. Assume that Q is a sound question, and that all the d-wffs in X are true. Then e-scenario Σ contains at least one path $\mathbf{s}$ such that:*

(1)   each d-wff of $\mathbf{s}$ is true,
(2)   each question of $\mathbf{s}$ is sound, and
(3) $\mathbf{s}$ leads to a true direct answer to Q.

The theorem states that if an initial question is sound and all the initial premises are true, then an e-scenario contains at least one *golden path*, which leads to a true direct answer to the initial question of the e-scenario (and that all the declaratives are true and all the auxiliary questions are sound). Intuitively, such an e-scenario provides a search plan which might be described as safe and finite—i.e. it will end up in a finite number of steps and it will lead to an answer to the initial question (cf. Wiśniewski, 2003, 2004a).

2.3.1.3. Entanglement

I will introduce one more concept here which will be useful for the following chapters. This is the concept of the *entanglement* of a d-wff in the context of an e-derivation. The concept is inspired by the entanglement of a formula in a synthetic inference introduced in (Urbański, 2005, p. 6). The main intuition here is to identify all the d-wffs from the set of initial premises of a given e-derivation that are relevant to the obtained direct answer to the initial question of this e-derivation.

I will start by introducing the $\Theta_{\mathbf{s}}$ set consisting of the initial premises of a given e-derivation $\mathbf{s}$ and all the direct answers collected to queries of this e-derivation.

Definition 35 *Let $\mathbf{s}$ be an erotetic derivation of a direct answer A to a question Q on the basis of a set of d-wffs X. Then:*

1. $\mathsf{d}\mathcal{Q}_{\mathbf{s}}$ *is the set of all the d-wffs of $\mathbf{s}$ which are direct answers to queries;*
2. $\Theta_{\mathbf{s}} = X \cup \mathsf{d}\mathcal{Q}_{\mathbf{s}}$.

Using the following corollary from (Wiśniewski, 2013b, p. 122):

Corollary 1 *If $\mathsf{Im}(Q, X \cup \{C\}, Q_1)$ and $X \models C$, then $\mathsf{Im}(Q, X, Q_1)$.*

We can say that all the relevant (declarative) information for obtaining the direct answer to the initial question in an e-derivation compresses to the d-wffs that are either elements of X or answers to the queries of this e-derivation.

Now we can introduce:

Definition 36 (Entanglement) *Let $\mathbf{s}$ be an erotetic derivation of a direct answer A to a question Q on the basis of a set of d-wffs X. Any set d-wffs Y (if existing) that satisfies the following conditions will be called entangled in a derivation of A:*

1. *all the elements of Y precede A in $\mathbf{s}$;*
2. *Y entails A;*
3. *$Y \subseteq \Theta_{\mathbf{s}}$;*
4. *there is no set Z of d-wffs such that all the elements of Z precede A in $\mathbf{s}$, Z entails A and $Z \subset Y$.*

Definition 37 (Entanglement of initial premises) *Let $\mathbf{s}$ be an erotetic deriva-tion of a direct answer A to a question Q on the basis of a set of d-wffs X. Let Y be a set of d-wffs entangled in the erotetic derivation of A. The set of initial declarative premises of $\mathbf{s}$ (Γ) entangled in a derivation of A is the following:*

$$\Gamma = X \cap Y$$

Example 5. Let us consider the e-scenario presented in Figure 2.7.

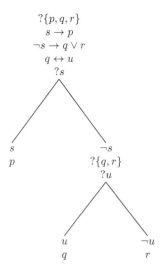

Fig. 2.7. Example of an e-scenario for the question $?\{p, q, r\}$ relative to the set of premises $\{s \to p, \neg s \to q \lor r, q \leftrightarrow u\}$ as an illustration of the idea of entanglement of d-wffs in e-derivation of the direct answer to the initial question

It consists of three paths, i.e. three e-derivations:

$$?\{p, q, r\}, s \to p, \neg s \to q \lor r, q \leftrightarrow u, ?s, s, p \tag{2.22a}$$

$$?\{p, q, r\}, s \to p, \neg s \to q \lor r, q \leftrightarrow u, ?s, \neg s, ?\{q, r\}, ?u, u, q \tag{2.22b}$$

$$?\{p, q, r\}, s \to p, \neg s \to q \lor r, q \leftrightarrow u, ?s, \neg s, ?\{q, r\}, ?u, \neg u, r \tag{2.22c}$$

For the leftmost path of the e-scenario (2.22a) the set of d-wffs entangled in the derivation of p is

$$\{s \to p, s\}$$

and the set of initial premises of our e-scenario entangled in the derivation of p is the following:

$$\{s \to p\}.$$

When it comes to path (2.22b) the set of d-wffs entangled in the derivation of q is:

$$\{q \leftrightarrow u, u\}.$$

and the set of initial premises of our e-scenario entangled in the derivation of q is the following:

$$\{q \leftrightarrow u\}.$$

For the path (2.22c) our set of entangled d-wffs is the following:

$$\{\neg s \to q \lor r, q \leftrightarrow u, \neg s, \neg u\}.$$

The set of the declarative premieses entangled in the derivation of r is:

$$\{\neg s \rightarrow q \vee r, q \leftrightarrow u\}.$$

2.3.2. Language $\mathcal{L}_{\mathbf{K3}}^{?}$

In this section I will introduce the language presented in (Leszczyńska-Jasion and Łupkowski, 2016). The language allows us to express questions with three possible answers: 'yes', 'no' or 'I don't know'. For the purpose of expressing the third answer Kleene's strong three-valued logic **K3** is used (see Urquhart, 2002). As the starting point we take the language $\mathcal{L}_{cpl}$ introduced in Section 2.3.1. The connectives are defined by the following truth-tables:

$\neg$	
1	0
1/2	1/2
0	1

$\wedge$	1	1/2	0
1	1	1/2	0
1/2	1/2	1/2	0
0	0	0	0

$\vee$	1	1/2	0
1	1	1	1
1/2	1	1/2	1/2
0	1	1/2	0

$\rightarrow$	1	1/2	0
1	1	1/2	0
1/2	1	1/2	1/2
0	1	1	1

$\leftrightarrow$	1	1/2	0
1	1	1/2	0
1/2	1/2	1/2	1/2
0	0	1/2	1

The language $\mathcal{L}_{cpl}$ is augmented with two additional unary connectives $\boxminus$ and $\boxplus$, and the new language is called $\mathcal{L}_{\mathbf{K3}}$. The set of wffs of $\mathcal{L}_{\mathbf{K3}}$ is defined as the smallest set containing the set of wffs of $\mathcal{L}_{cpl}$ and such that if A is a wff of $\mathcal{L}_{cpl}$, then

(i) $\boxminus(A)$ is a wff of $\mathcal{L}_{\mathbf{K3}}$ and
(ii) $\boxplus(A)$ is a wff of $\mathcal{L}_{\mathbf{K3}}$.

It should be stressed that the new connectives never occur inside the wffs of $\mathcal{L}_{cpl}$ and they cannot be iterated.

The intended reading of the new connectives is the following:

(i) $\boxminus A$—an agent a *cannot* decide if it is the case that A using the knowledge base D.
(ii) $\boxplus A$—an agent a *can* decide if it is the case that A using the knowledge base D.

The connectives are characterised by the following truth-table:

A	$\boxminus A$	$\boxplus A$
1	0	1
1/2	1	0
0	0	1

Thus semantically speaking, the new connectives reduce the set of possible values to the classical ones. Let us also observe that if we are allowed to put negation in front of $\boxminus$, then the second connective could be introduced by the following definition:

$$\boxplus A =_{df} \neg \boxminus A.$$

We will define language $\mathcal{L}^?_{\mathbf{K}3}$, built upon $\mathcal{L}_{\mathbf{K}3}$, in which questions may be formed. The vocabulary of $\mathcal{L}^?_{\mathbf{K}3}$ contains the vocabulary of $\mathcal{L}_{\mathbf{K}3}$ (thus also the connectives $\boxplus$, $\boxminus$) and the signs: ?,{,}. As in the classical case, by the declarative well-formed formulas of $\mathcal{L}_{\mathbf{K}3}$ (d-wffs) we mean the well-formed formulas of $\mathcal{L}_{\mathbf{K}3}$. The notion of an e-wff (a question) of $\mathcal{L}^?_{\mathbf{K}3}$ is also defined as in the classical case (see (2.2), page 30), this time, however, the direct answers to a question are formulated in $\mathcal{L}_{\mathbf{K}3}$. We apply all the previous notational conventions.

We are interested in a specific category of *ternary* questions, which may be viewed as the counterparts of simple yes-no questions, provided with the third possible answer "it is not known whether". Thus a ternary question will be represented in language $\mathcal{L}^?_{\mathbf{K}3}$ as follows:

$$?\{A, \neg A, \boxminus A\} \tag{2.23}$$

Expressions of the form (2.23) will be called *ternary questions of $\mathcal{L}^?_{\mathbf{K}3}$*. For the sake of simplicity, I will represent them as '$?A$'. If A in schema (2.23) is a propositional variable, then (2.23) is called an *atomic ternary question of $\mathcal{L}^?_{\mathbf{K}3}$* or, more specifically, an *atomic ternary question of $\mathcal{L}^?_{\mathbf{K}3}$ based on the propositional variable A*. I adapt here the same notational convention as for binary questions (see (2.4), page 30). In what follows it will be clear from the context whether $?A$ refer to a binary or a ternary question.

The following notation:

$$? \pm \boxminus |A, B| \tag{2.24}$$

refers to a question of the form:

$$?\{A {\wedge} B, A {\wedge} \neg B, A {\wedge} \boxminus B, \neg A {\wedge} B, \neg A {\wedge} \neg B, \neg A {\wedge} \boxminus B, \boxminus A {\wedge} B, \boxminus A {\wedge} \neg B, \boxminus A {\wedge} \boxminus B\} \tag{2.25}$$

(2.25) is a ternary counterpart of binary conjunctive questions (recall (2.5), page 31).

We will use the general setting of MiES here.

Definition 38 (Partition of $\mathcal{L}_{\mathbf{K}3}$) *Let $\mathcal{D}_{\mathcal{L}^?_{\mathbf{K}3}}$ be the set of d-wffs of language $\mathcal{L}^?_{\mathbf{K}3}$. By a partition of language $\mathcal{L}^?_{\mathbf{K}3}$ we mean an ordered pair $\mathsf{P} = \langle \mathsf{T_P}, \mathsf{U_P} \rangle$ such that:*

- $\mathsf{T_P} \cap \mathsf{U_P} = \emptyset$
- $\mathsf{T_P} \cup \mathsf{U_P} = \mathcal{D}_{\mathcal{L}^?_{\mathbf{K}3}}$

By a *partition of the set* $\mathcal{D}_{\mathcal{L}^?_{\mathbf{K}3}}$ we mean a partition of language $\mathcal{L}^?_{\mathbf{K}3}$. If for a certain partition P and a d-wff A, $A \in \mathsf{T_P}$, then we say that A is *true in partition* P, otherwise, A is *untrue in* P. What is essential for the semantics of $\mathcal{L}^?_{\mathbf{K}3}$ is the notion of a **K3**-admissible partition. First, we define the notion of a **K3**-*assignment* as a function $VAR \longrightarrow \{0, \frac{1}{2}, 1\}$. Next, we extend **K3**-assignments to **K3**-*valuations* according to the truth-tables of **K3**. Now we are ready to present:

Definition 39 (Admissible partition of $\mathcal{L}_{\mathbf{K}3}$) *We will say that partition* P *is* **K3**-admissible *provided that for a* **K3**-*valuation* V, *the set* $\mathsf{T_P}$ *consists of formulas true under* V *and the set* $\mathsf{U_P}$ *consists of formulas which are not true under* V.

As for $\mathcal{L}^?_{cpl}$ we can now introduce the notions of sound and safe questions.

Definition 40 (Soundness) *A question* Q *is called sound under a partition* P *provided that some direct answer to* Q *is true in* P.

Definition 41 (Safety) *We will call a question* Q *safe, if* Q *is sound under each* **K3**-*admissible partition.*

Note that in the three-valued setting a polar question (i.e. question of the form $?\{A, \neg A\}$) is not safe. However, each ternary question of $\mathcal{L}^?_{\mathbf{K}3}$ is safe.

We will make use of the notion of a multiple-conclusion entailment[7], which denotes a relation between sets of d-wffs generalising the standard relation of entailment (see page 32).

Definition 42 (Multiple-conclusion entailment in $\mathcal{L}^?_{\mathbf{K}3}$) *Let* X *and* Y *be sets of d-wffs of language* $\mathcal{L}^?_{\mathbf{K}3}$. *We say that* X mc-*entails* Y *in* $\mathcal{L}^?_{\mathbf{K}3}$, *in symbols* $X \Vdash_{\mathcal{L}^?_{\mathbf{K}3}} Y$, *iff there is no* **K3**-*admissible partition* $\mathsf{P} = \langle \mathsf{T_P}, \mathsf{U_P} \rangle$ *of* $\mathcal{L}^?_{\mathbf{K}3}$ *such that* $X \subseteq \mathsf{T_P}$ *and* $Y \subseteq \mathsf{U_P}$.

Now we may introduce:

Definition 43 (Entailment in $\mathcal{L}^?_{\mathbf{K}3}$) *Let* X *be a set of d-wffs and* A *a single d-wff of* $\mathcal{L}^?_{\mathbf{K}3}$. *We say that* X *entails* A *in* $\mathcal{L}^?_{\mathbf{K}3}$, *in symbols* $X \models_{\mathcal{L}^?_{\mathbf{K}3}} A$, *iff* $X \Vdash_{\mathcal{L}^?_{\mathbf{K}3}} \{A\}$, *that is, iff for each* **K3**-*admissible partition* P *of* $\mathcal{L}^?_{\mathbf{K}3}$, *if each formula from* X *is true in* P, *then* A *is true in* P.

2.3.2.1. Erotetic implication revisited

Now we are ready to introduce the notion of erotetic implication in $\mathcal{L}^?_{\mathbf{K}3}$. It is worth noticing that the underlying intuitions here are exactly the same as in the case of Definitions 24 and 25 (page 33).

---

[7] See (Shoesmith and Smiley, 1978).

Definition 44 (Erotetic implication in $\mathcal{L}_{\mathbf{K}3}^?$) *Let Q and Q^* stand for questions of $\mathcal{L}_{\mathbf{K}3}^?$ and let X be a set of d-wffs of $\mathcal{L}_{\mathbf{K}3}^?$. We will say that Q $\mathcal{L}_{\mathbf{K}3}^?$-implies Q^* on the basis of X, in symbols $\mathbf{Im}_{\mathcal{L}_{\mathbf{K}3}^?}(Q, X, Q^*)$, iff*

1. *for each $A \in \mathbf{d}Q$, $X \cup \{A\} \mathrel{\|\!\!\!=}_{\mathcal{L}_{\mathbf{K}3}^?} \mathbf{d}Q^*$, and*
2. *for each $B \in \mathbf{d}Q^*$, there is a non-empty proper subset Y of $\mathbf{d}Q$ such that $X \cup \{B\} \mathrel{\|\!\!\!=}_{\mathcal{L}_{\mathbf{K}3}^?} Y$.*

Definition 45 (Pure erotetic implication in $\mathcal{L}_{\mathbf{K}3}^?$) *Let Q and Q^* stand for questions of $\mathcal{L}_{\mathbf{K}3}^?$. We will say that Q $\mathcal{L}_{\mathbf{K}3}^?$-purely implies Q^*, in symbols $\mathbf{Im}_{\mathcal{L}_{\mathbf{K}3}^?}(Q, Q^*)$, iff*

1. *for each $A \in \mathbf{d}Q, \mathrel{\|\!\!\!=}_{\mathcal{L}_{\mathbf{K}3}^?} \mathbf{d}Q^*$, and*
2. *for each $B \in \mathbf{d}Q^*$, there is a non-empty proper subset Y of $\mathbf{d}Q$ such that $B \mathrel{\|\!\!\!=}_{\mathcal{L}_{\mathbf{K}3}^?} Y$.*

Let us consider examples of e-implication in $\mathcal{L}_{\mathbf{K}3}^?$.

$$\mathsf{Im}(?\{A \wedge B, \neg(A \wedge B), \boxminus(A \wedge B)\}, ? \pm \boxminus|A, B|) \tag{2.26}$$

$$\mathsf{Im}(? \pm \boxminus|A, B|, ?\{A, \neg A, \boxminus A\}) \tag{2.27}$$

$$\mathsf{Im}(? \pm \boxminus|A, B|, ?\{B, \neg B, \boxminus B\}) \tag{2.28}$$

$$\mathsf{Im}(?\{\neg A, \neg\neg A, \boxminus\neg A\}, ?\{A, \neg A, \boxminus A\}) \tag{2.29}$$

$$\mathsf{Im}(?\{A \vee B, \neg(A \vee B), \boxminus(A \vee B)\}, ? \pm \boxminus|A, B|) \tag{2.30}$$

$$\mathsf{Im}(?\{A \rightarrow B, \neg(A \rightarrow B), \boxminus(A \rightarrow B)\}, ? \pm \boxminus|A, B|) \tag{2.31}$$

$$\mathsf{Im}(?\{A \leftrightarrow B, \neg(A \leftrightarrow B), \boxminus(A \leftrightarrow B)\}, ? \pm \boxminus|A, B|) \tag{2.32}$$

$$\mathsf{Im}(?\{A, \neg A, \boxminus A\}, \{A \leftrightarrow B \wedge C, \boxminus B\}, ?\{C, \neg C, \boxminus C\}) \tag{2.33}$$

2.3.2.2. Erotetic search scenario revisited

When it comes to e-scenarios in $\mathcal{L}_{\mathbf{K}3}^?$ all the basic notions are defined as for $\mathcal{L}_{cpl}^?$. An e-scenario is understood as a finite family of e-derivations (see Definition 26). A query of an e-scenario is a query of e-derivation (see Definition 28), i.e. a query of a path of the e-scenario. Exemplary e-scenarios of $\mathcal{L}_{\mathbf{K}3}^?$ are presented in Figures 2.8., 2.9. and 2.10.

As in the classical case (see page 39) we arrive at:

Theorem 3 (Golden Path Theorem). *Let Σ be an e-scenario for a question Q (of language $\mathcal{L}^{?}_{\mathbf{K3}}$) relative to a set X of d-wffs (of $\mathcal{L}^{?}_{\mathbf{K3}}$). Let* P *be a K3-admissible partition of language $\mathcal{L}^{?}_{\mathbf{K3}}$ such that Q is sound in* P *and all the d-wffs in X are true in* P*. Then the e-scenario Φ contains at least one path* s *such that:*

1. *each d-wff of* s *is true in* P*,*
2. *each question of* s *is sound in* P*, and*
3. s *leads to a direct answer to Q which is true in* P*.*

The proof of the theorem is presented in (Leszczyńska-Jasion and Łupkowski, 2016, p. 67).

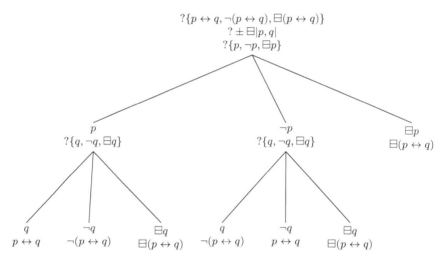

Fig. 2.8. E-scenario (pure) for question $?\{p \leftrightarrow q, \neg(p \leftrightarrow q), \boxminus(p \leftrightarrow q)\}$. Compare with pure e-scenario for language $\mathcal{L}^{?}_{cpl}$ presented in Figure 2.6.

An introduction of ternary questions and e-scenarios for them allows us to express certain pragmatic features of the e-scenarios. Let us imagine that an agent a wants to establish whether A is the case. The agent knows that: $A \leftrightarrow B \wedge C$, but knows nothing about B. We may now imagine that a solves his/her problem according to the e-scenario presented in Figure 2.10. (as can be observed, a's premise and the fact that $\boxminus B$ are incorporated in the initial premises of the e-scenario).

In this example $\boxminus B$ might be treated as an information gap. However, it can be observed that there is one possible course of events that will lead to the answer to the initial question despite a lack of knowledge about B—namely

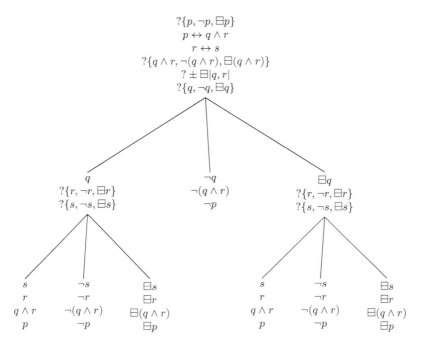

Fig. 2.9. Example of an e-scenario for the question $?\{p, \neg p, \boxminus p\}$ relative to the set of premises $\{p \leftrightarrow q \wedge r, r \leftrightarrow s\}$. Compare with the e-scenario for language $\mathcal{L}^?_{cpl}$ presented in Figure 2.4.

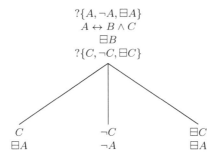

Fig. 2.10. Schema of an e-scenario for the question $?\{A, \neg A, \boxminus A\}$ relative to the set of premises $A \leftrightarrow B \wedge C, \boxminus B$. Notice that the lack of knowledge about B is expressed in the premises

the case, where the answer to the question $?\{C, \neg C, \boxminus C\}$ is negative (then the answer to the initial question is also negative). We may say that the proposed e-scenario offers three cognitive situations (from most to least preferable):

- A 'maximal' cognitive situation is represented by the path going through the answer $\neg C$, because it leads to $\neg A$, i.e. a definite answer to the initial question.
- A 'minimal' one is reflected by the path which goes through the answer C, as in this situation the questioning process ends up with some knowledge gains (despite the fact that we did not manage to solve the initial problem, we know C).
- A 'zero knowledge' situation is represented by the third path going through $\boxminus C$, because it finishes the questioning process without any knowledge gains.

The e-scenario presented in Figure 2.8. consists of four maximal paths, two minimal paths (leading through $\boxminus q$) and one zero-knowledge path (leading through $\boxminus p$—the rightmost path of the e-scenario). The e-scenario in Figure 2.9. consists of five maximal paths. The interesting cases may be observed for the rightmost sub-tree (starting from $\boxminus q$). Here we have two maximal paths leading to the answer to the initial question (respectively p and $\neg p$) despite the lack of knowledge about q. The e-scenario also has one minimal path (leading through q, $\boxminus s$, $\boxminus r$ to $\boxminus p$) and one zero-knowledge path (the rightmost path of the e-scenario).

2.4. Summary

This chapter is the backbone of the book and is the key to the ideas presented in the later chapters. Here I have introduced basic concepts of Inferential Erotetic Logic. This short introduction to IEL allows for a better understanding of the way questions are formalised within this approach. Some crucial ideas and notions were introduced here (question, direct answer, erotetic implication and erotetic search scenario). I have also introduced formal languages ($\mathcal{L}_{cpl}^{?}$ in Section 2.3.1 and $\mathcal{L}_{\mathbf{K}3}^{?}$ in Section 2.3.2) that are used further on in this book. A reader will also find a brief overview of a wide range of IEL applications (from proof-theory to empirical research).

Chapter 3

Dependency of questions in natural language dialogues

This chapter is an attempt to provide an answer to the question: how does natural language data relate to the picture of questioning and answering presented in Chapter 2? The vast majority of data analysed in this chapter (especially in 3.1) comes from a larger corpus study (Łupkowski and Ginzburg, 2013) aimed at investigating the phenomenon of answering questions with questions (*query responses*).

The study covered:

1. The Basic Electricity and Electronics Corpus (BEE) (Rosé et al., 1999): tutorial dialogues from electronics courses.
2. The British National Corpus (BNC): large balanced corpus.
3. The Child Language Data Exchange System (CHILDES) (MacWhinney, 2000): adult-child conversations.
4. The SRI/CMU American Express dialogues (AMEX) (Kowtko and Price, 1989): conversations with travel agents.

The study sample consists of 1.466 query/query response pairs. As an outcome the following query responses (q-responses) taxonomy was obtained: (1) CR: clarification requests; (2) DP: dependent questions, i.e. cases where the answer to the initial question depends on the answer to a q-response (see Definition 2 in Chapter 1.1.1); (3) MOTIV: questions about an underlying motivation behind asking the initial question; (4) NO ANSW: questions aimed at avoiding answering the initial question; (5) FORM: questions considering the way of answering the initial question; (6) QA: questions with a presupposed answer, (7) IGNORE: responses ignoring the initial question—for more details see (Łupkowski and Ginzburg, 2013, p. 355).

The most interesting class of query responses from my perspective are dependent questions. My claim here is that dependency of questions can be efficiently analysed in terms of erotetic implication.

Let us get back to the intuition behind the notion of question dependence. It states that whenever an answer to question Q_2 (i.e. the one given as a response) is reached then some information about answer to Q_1 (i.e. the initial question) becomes available. This allows one to say that Q_2 is a relevant response to Q_1 (q-specific utterance—see Definition 1 in Chapter 1.1.1) as illustrated by the following example (5):

(5)   A: Any other questions?

       B: Are you accepting questions on the statement of faith at
             this point?
             $[F85, 70\text{-}71]^1$
             [*Whether more questions exist depends on whether you are accepting*
             questions on the statement of faith at this point.]

It may be observed that in (5) the answer to A's question will depend on the answer to the question asked by B. The information about whether questions on the statement of faith are accepted at this point provides the information about the question of whether more questions exist.

As the aforementioned corpus study revealed, dependent questions are common as q-responses. In more generally oriented corpora such as the BNC and CHILDES dependent questions provided as answers are the second largest class among the recognised types of query responses. When corpora with goal-directed dialogues are considered (BEE, AMEX) dependent questions are the most frequent category of query responses. For a more detailed discussion see (Łupkowski and Ginzburg, 2013).

Let us now focus our attention on the BEE data and take a closer look at the dependent q-responses in the context of tutorial dialogues. Understanding such a dependency between questions gives us an insight on the sub-questionhood relation (i.e. replacing the issue raised by the initial question by an easier to answer sub-issue raised by a dependent q-response), which plays an important role in these fairly cooperative and goal-directed tutorial dialogues.

The choice of tutorial dialogues domain is intuitive when we think about question processing and questions used to facilitate the answer search. As Van-Lehn et al. (2006, p. 1) put it: "A tutor often faces the decision of whether to just tell the student an explanation or to try to elicit the explanation from the student via a series of questions."

In the BEE study, 45 query/query responses pairs were identified. As already mentioned, the dependent q-responses are the most common ones used in the sample—28 cases, which constitutes 62% of the sample. The structure of the q-responses identified in the BEE study is presented in Figure 3.1.

What is important, 19 of the initial questions and 16 of the dependent q-responses from the sample are simple yes/no questions (further on I shall represent an initial question by Q_1 and q-response as Q_2). This might be explained by the reference to the main task of dependent questions—they should facilitate providing an answer to the initial question. What is more, one may think of dependent query responses in terms of leading to a subsequent answer to Q_2 and consequently to an answer to Q_1. Our expectation here would be that

---

[1] This notation indicates the BNC file (F85) together with the sentence numbers (70–71).

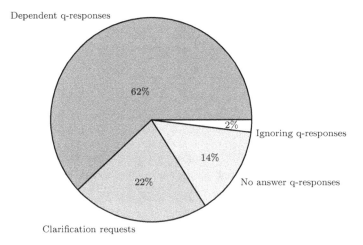

Fig. 3.1. Types of query responses identified in BEE

answering Q_2 should lead to answering Q_1. The issue raised as the main question should be resolved as parties of a dialogue cooperate to achieve this goal. This expectation should be reflected in the percentage of answered questions (initial and q-responses). The percentage of the answered initial questions and the percentage for the answered q-responses should be high and the numbers should be similar. Analysis of the BEE data reveals that all pairs of initial question and dependent q-response were provided with answers (for Q_1 and for Q_2).

Our expectation gets its confirmation in analysis of other goal-directed dialogues (see Łupkowski and Ginzburg, 2013). For the AMEX corpus we have 96.43% answered initial questions; and exactly the same percentage—96.43%—of answered q-responses. We can compare these numbers with more generally oriented dialogues. For the BNC we have: 62.96% for Q_1 and 76.85% for Q_2, while for the CHILDES corpus we have: 45.45% for Q_1 and 72.73% for Q_2 respectively—so the numbers are still high (lower result for Q_1 in the case of CHILDES may be explained by the nature of the data—when it comes to interaction with children maintaining attention and topic continuity is more difficult).

We can also check what types of initial questions lead to what types of query responses. In the BEE case it is that in 44% of cases a simple yes/no question was given as a response to a simple yes/no question. 25.93% of cases lead from a simple yes/no initial question to an open q-response. As for open initial questions 14.81% of cases lead to an open q-response, while 14.81% lead to simple yes/no q-responses. The summary is presented in Table 3.1.

Table 3.1. Dependent questions in BEE—response pattern

Q_1	leads to	Q_2
yes/no	yes/no	44.44%
yes/no	open	25.93%
open	yes/no	14.81%
open	open	14.81%

The discussed data reveals the following pattern of relations between questions in tutorial dialogues (which applies to a large extent also to more general dialogues):

- the most frequent category of query responses are dependent ones, which facilitate the search for an answer to the initial question;
- simple yes-no questions are often used;
- in most cases when a simple yes/no question is the initial one it will be followed by a simple yes/no question-response;
- both the question and its query response are answered in all the analysed cases—this suggest the high level of cooperation between dialogue participants.

3.1. Dependency of questions and erotetic implication

In this section I will consider how e-implication might serve as a tool in analysing certain dialogue moves. As discussed in the first chapter of this book I will not treat logical concepts as the ultimate explanation of linguistic phenomena (i.e. that people should process questions according to IEL). My approach here is rather that logic provides a very useful normative yardstick to study, describe and analyse these phenomena.

The analysis presented in this section use the language $\mathcal{L}^?_{cpl}$ introduced in Chapter 2.3.1.

Let us now go through examples of answering questions with questions retrieved from the language corpora. We will start with the following example:

(6)   A: Do you want me to <*pause*> push it round?
       B: Is it really disturbing you?
             [*FM1, 679–680*]

In order to explain the rationale behind B's dialogue move observed in (6) I will use the notion of e-implication. What is the reason behind B's decision to response with a question to A's question in this case? Why would B not simply provide an answer to A's question? We reconstruct this case in a manner described in Chapter 2.2. B has a piece of information that allows for the process of A's question and produces a dependent auxiliary question. When we investigate the example we may notice that this is the case—B's question-answer in this case is certainly a dependent question (cf. *Whether I want you to push it depends on whether it really disturbs you*). This additional information is something that B probably accepts in this dialogue situation (but is not explicitly given by B). One of the possibilities in this case might be: 'I want you to push it round if and only if it is disturbing you'. If we accept his premise, our example will appear as (7). We may consider this additional information to play the role of enthymematic premise in the analysed erotetic inference.

(7)   A: Do you want me to <*pause*> push it round?
       B: *I want you to push it round iff it is disturbing you.*
       B: Is it really disturbing you?

In the situation presented above A's question might be interpreted as an expression of a problem/issue to solve. In other words, after the public announcement of the question "Do you want me to <*pause*> push it round?" it becomes a problem to be solved by B. We may say, that this question becomes B's initial question that has to be processed. Interpreted this way it allows us to grasp the rationale behind B's dialogue move in the example. Let us take a look at the logical structure of (7). Now we are focused only on B's side, because the question about pushing the thing around is B's initial question. In (7) $?p$ represents B's initial questions (introduced in dialogue by A): "Do I want to push it round?". Formula $p \leftrightarrow q$ represents the piece of information used by B to decompose $?p$. After the initial question is processed B comes to the question $?q$ ("Is it really disturbing you?"), which is later publicly announced (asked) to A.

(8)   B: $?p$
       B: $p \leftrightarrow q$
       B: $?q$

How can we evaluate this dialogue move? We may conclude that B's dialogue move is well-motivated from the normative point of view, because the following holds:

$$\mathsf{Im}(?p, p \leftrightarrow q, ?q) \tag{3.1}$$

B's q-response at which B arrives after the initial question processing and later asks it in this case is e-implied (on the basis of the introduced declarative premise). This is to say that the motivation behind this dialogue move is explained by the e-implication. Let us remember the core intuitions behind e-implication. For e-implication the implied question obtained as a result of the decomposition process should have certain characteristics. First of all, it should stay on the main topic. In other words, no random questions should appear here. However the main characteristic that we aim at here is that the answer provided to the auxiliary question should be at least a partial answer

to the initial question. It should bring our agent closer to solving the initial
problem. Summing up, we can perceive the discussed process of replacing one
question with another (simpler) one as a justified step from the problem-solving
perspective. Such an explanation is in line with the picture of questioning observed for the corpus data presented in the introduction to this chapter.

Let us observe that example (5) on page 50 has the same structure as a
dialogue part presented in (6). For the case of (5)—in the schema (8)—p would
stand for "Do I have any other question?"; $p \leftrightarrow q$ for "I would have a question if
and only if questions on the statement of faith are accepted at this point." and
$?q$—"Are such questions accepted?". This sends us to the implication presented
in (3.1).

We may also analyse example (9) in a similar manner.

(9)   A: Pound, five ounces, do you want a wee bit more?
       B: Er, erm <*pause*> can I see it?
             [*KC0, 1812–1813*]

Similarly to the previous case, we have to introduce a premise accepted by B
in this situation because it is not expressed in the dialogue. We may say that
a plausible one might be: 'I want more if and only if I can see it, and I am
satisfied with it'. Then our example will have the structure presented in (10a).

(10) a. A: Pound, five ounces, do you want a wee bit more?
          B: *I want more iff I can see it and I am satisfied with it.*
          B: Er, erm <*pause*> can I see it?
      b. B: $?p$
          B: $p \leftrightarrow q \wedge r$
          B: $?q$

The logical structure of the inference in question (on B's side) is presented
in (10b). Here $?p$ stands for "Do I want a wee bit more?"; $p \leftrightarrow q \wedge r$ expresses
our introduced premise, and $?q$ represents the auxiliary question which is a
result of the initial question processing with respect to the premise—that later
is asked to A.

And analogically to the previous case—see (6), because the following holds:

$$\mathsf{Im}(?p, p \leftrightarrow q \wedge r, ?(q \wedge r)) \tag{3.2a}$$
$$\mathsf{Im}(?(q \wedge r), ? \pm |q, r|) \tag{3.2b}$$
$$\mathsf{Im}(? \pm |q, r|, ?q) \tag{3.2c}$$

we may say that in this case B's answer is e-implied by the initial question on
the basis of the introduced declarative premise.

Let us take a look at yet another example (11a).

(11) a. A: Shut up <*pause*> can I have a sarnie?

B: Do you want one?
[*KBR, 1791–1792*]

b. B: Shut up *<pause>* can I have a sarnie?
  B: *If you want a sarnie <u>then</u> you can have it.*
  B: Do you want one?

c. A: $?p$
  B: $q \to p$
  B: $?q$

The premise accepted by B is presented in (11b) and the logical structure of the example is explicated in (11c). The logical facts (3.3a) and (3.3b) serve as the normative explanation of the dialogue move our example.

$$\mathsf{Im}(?p, q \to p, ?\{p, \neg p, q\}) \tag{3.3a}$$

$$\mathsf{Im}(?\{p, \neg p, q\}, ?q) \tag{3.3b}$$

At this point our approach seems to have one serious problem. In the analysed examples (6), (9) and (11a) we were forced to introduce some extra declarative premises accepted by one of the dialogue participants (since they were not expressed in the dialogue).[2] Fortunately, there are cases, where a dialogue is rich enough to reveal such premises (or at least makes it possible to convincingly reconstruct them). Let us take a look at the example of a conversation between the tutor (T) and the student (S):

(12)   TUTOR: True—but if they are all moving at the same speed, will anyone actually ever run into anyone else?
   STUDENT: Are they all going in the same direction?
   STUDENT: If they are they would not run into each other but if they are going in all kinds of different directions they might.
   [*BEE(F), stud25*][3]

In this example, the premises accepted by S are clearly expressed (see the third utterance in 12). Let us remember that in our interpretation we take T's question as a formulation of the initial problem for S. We may say that after the public announcement of the question by T it becomes S's initial question. This question has to be processed with respect to the knowledge at his/her

---

[2] To some extent this is a result of the language data used—i.e. language corpora. In this case we work on given material. One can imagine a more flexible way of collecting interesting data for the purpose of question processing research, such as e.g. a game with a purpose (see Łupkowski, 2011b).

[3] This notation indicates BEE sub-corpus (F—Final Experiment) and the file number (stud25). Unfortunately no sentence numbering is available for the BEE corpus.

disposal. In (12) pieces of knowledge used by the Student to analyse the initial question are clearly expressed. Consequently our example might be presented as in (13a), or—in a simpler way—like in (13b):

(13) a. INITIAL Q.: $?p$
          FACTS:
          $$q \rightarrow \neg p$$
          $$\neg q \rightarrow p$$
          AUXILIARY Q.: $?q$
      b. INITIAL Q.: $?p$
          FACT: $p \leftrightarrow \neg q$
          AUXILIARY Q.: $?q$

Given this, and the following logical fact:

$$\mathsf{Im}(?p, p \leftrightarrow \neg q, ?q) \tag{3.4}$$

we may say that the question introduced by S as a response to T's question was e-implied (on the basis of the expressed declarative premises).

The erotetic inference step might be interpreted as follows. Tutor announces the initial question "True—but if they are all moving at the same speed, will anyone actually ever run into anyone else?". Our student does not know the answer to this initial question. The initial question is being processed with the use of the student's knowledge, namely the fact that: "If they are they would not run into each other but if they are going in all kinds of different directions they might.". Notice that this piece of information is revealed by S in order to explain why the following question-response appears in the dialogue: "Are they all going in the same direction?" When S obtains an answer to this question he/she will be able to answer the initial one. Thus one may say that the question-response of the student is a justified step in this dialogue.

Let us consider another example taken from the BEE corpus.

(14)   TUTOR: Can you tell me how current would flow? Would it flow only through the wire?
          Or only through the meter?
          Or through both?
          Or through neither?
          STUDENT: I don't know.
          TUTOR: Is there any reason why it wouldn't flow through the wire?
          STUDENT: I don't know.
          Would the leads prevent or obstruct it in some way?
          TUTOR: No.
          If you know that the leads do not obstruct the current flow in any way, can you answer my question?

STUDENT: I guess I'd have to say yes.
The current would flow through both.
TUTOR: Good.
[*BEE(F), stud38*]

We can simplify this conversation to show its structure more clearly:

(15) a. TUTOR: Can you tell me how current would flow? Would it flow only through the wire? Or only through the meter? Or through both? Or through neither?
      b. STUDENT: Would the leads prevent or obstruct it in some way?
      c. TUTOR: No.
      d. STUDENT: The current would flow through both.

(15a) is the initial problem/question, while (15d) is an answer to this question. What we want to explain is how the auxiliary question (15b) appeared. What is important in order to provide this explanation is what is really asked by the Tutor: *If you know that the leads do not obstruct the current flow in any way, can you answer my question?* This statement points to the shared knowledge of T and S, i.e. that the current will flow through the wire and the meter *iff* the leads do not obstruct the current in any way. There are presumably other premises assumed by the dialogue participants, but this one is clearly stated in the conversation.

The initial problem is expressed by a conjunctive question: $?\pm|A,B|$. The premise revealed in the dialogue is of the form: $A \wedge B \leftrightarrow \neg C$ (where C—'leads obstruct the current flow').

Since the following holds: $\mathsf{Im}(?\pm|A,B|, A \wedge B \leftrightarrow \neg C, ?C)$, we may say that the Student's question we are interested in was e-implied by the initial question and the premise.

Our next example (16) brings certain new issues into our analysis. It is also retrieved from the BEE corpus.

(16)   TUTOR: When there is a non-zero voltage reading, there is a difference in charge.
        If there is a zero voltage reading, there is no difference in charge.
        In a switch, is there any voltage being "created" or "used up"?
        STUDENT: no
        TUTOR: Right, so would you expect there to be a difference in charge across it?
        STUDENT: no
        TUTOR: So what would you expect the voltage reading to be?
        STUDENT: 0
        TUTOR: Right.
        [*BEE(F), stud37*]

In this case we have to take into account the broader context of the dialogue. What is interesting here, is that questions are used in the context of teaching. This might illustrate how question processing is used to facilitate the learning process.

Let us try to explain how the Tutor arrives at the question 'In a switch, is there any voltage being "created" or "used up"?' and introduces it into the dialogue. We need to use knowledge introduced by T earlier in this dialogue (17a), which might be summarised as in (17b).

(17) a. TUTOR: You can get a voltage reading across a source (list a battery) or a load (like a lightbulb) because there is a difference in charge between the two ends.

This is different from just polarity—it's a difference caused by there either being voltage "created" (as in a source) or "used up" as in a load.

You just have to make sure that where you attach the leads, there is a difference in charge between where the red lead is attached and where the black lead is attached.

[*BEE(F), stud37*]

b. TUTOR: *There is a difference in charge if and only if voltage is "created" or "used up".*

As a consequence we have three premises relevant to the considered example:

(1)   When there is a non-zero voltage reading, there is a difference in charge.
(2)   If there is a zero voltage reading, there is no difference in charge.
(3)   There is a difference in charge if and only if voltage is "created" or "used up".

Premises (1) and (2) have the following structure: (1) $\neg p \rightarrow q$ and (2) $p \rightarrow \neg q$—so (1) and (2) jointly are equivalent to $p \leftrightarrow \neg q$. Premise (3) has the following structure: $q \leftrightarrow r$. It is important that we also consider initial question which is being considered by T and S in our exemplary dialogue. The initial question (problem) considered by T and S in this dialogue might be reconstructed as: 'Is it the case that there is a zero voltage reading?'

How should we explain the appearance of the question about voltage being "created" or "used up"? T processes the initial question by making use of the pieces of information provided to S. Afterwards T asks our auxiliary question. It becomes well-motivated, when we consider the following:

$$\mathsf{Im}(?p, p \leftrightarrow \neg q, ?q) \qquad\qquad (3.5\text{a})$$

$$\mathsf{Im}(?q, q \leftrightarrow r, ?r). \qquad\qquad (3.5\text{b})$$

It is worth stressing that there is a certain strategy (or hidden agenda) behind T's auxiliary question. T's aim is to check whether S understands the issue expressed by the initial question—see the next section (3.2). This becomes even more clear when we take a closer look at the way the conversation goes further. The answer to the question about voltage $(?r)$ is negative. This (together with the premise $r \leftrightarrow q$) provides a negative answer to the question about q. And since we have $p \leftrightarrow \neg q$, the answer to the initial question is p, i.e. that there is a zero voltage reading. If we wished to present this case in the form of a tree, which would present all possible courses of events we would obtain an e-scenario like the one presented in Figure 3.2. The initial question is the root and the leaves are the answers to the initial questions. Two branches splitting after the question $?r$ represent two possible courses of events in the dialogue—depending on the answer to $?r$. In case under question the rightmost branch is activated by the negative answer to $?r$.

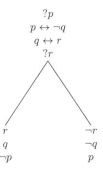

Fig. 3.2. E-scenario for the dialogue in example (16)

One more remark is in order here. It seems that the Tutor is asking questions: 'Right, so would you expect there to be a difference in charge across it?' and 'So what would you expect the voltage reading to be?' in order to make sure that the Student really understands the consequences of a negative answer to the question about voltage. This is indicated by the phrase *would you expect* present in both of these questions.

3.2. Questioning strategies

In the previous section, I analysed how a dialogue participant arrives at a question-response and introduces it into a dialogue on the basis of the initial question and a declarative premise(s). The aim was to grasp the idea of question dependency and of using questions as dialogue moves in the light of a normative notion of e-implication. In this section we will focus on something different. IEL also makes it possible to model broader parts of dialogues seen from an interrogator's perspective in order to reveal his/her questioning strategy. This approach is presented in (Łupkowski, 2010, 2011a) and in (Urbański and Łupkowski, 2010b). In this section I would like to use it to analyse dialogue examples retrieved from the BEE corpus.

An e-scenario might be viewed as providing a search plan for an answer to the initial question. This plan is relative to the premises a questioner has, and leads through auxiliary questions (and the answers to them) to the initial question's answer. The key feature of e-scenarios is that auxiliary questions appear in them on the condition that they are e-implied. Thus we may use e-scenarios to give some insights into questioning strategies used in dialogues. This approach is efficient for contexts where a questioner wants to obtain an answer to the initial question, which should not be asked directly (as e.g. in the Turing test situation, where asking a direct question 'Are you a human or a machine' would be fruitless as a satisfactory solution to the problem of an agent's identification).[4] To obtain an answer to the initial question, the questioner usually asks a series of auxiliary questions in such situations. Answers to these questions build up to an answer to the initial one. It is easy to imagine such a context in real life situations, like for example while teaching, when we want to check if our student really understands a given problem. In what follows I will analyse such examples retrieved from the BEE corpus.

First let us consider example (18).

(18)   TUTOR: No, you don't have to hook the leads directly to the battery.
         The important thing is that you observe polarity.
         Do you know what the rule for observing polarity is?
         STUDENT: Yes I do.
         TUTOR: Can you tell me?
         STUDENT: It's when the positive end of the meter is connected to the positive end of the circuit and the same for the negative end
         TUTOR: Right, but do you know how to determine what is the positive side of the circuit and what is the negative side? If the red lead were hooked up to tab #2, which tab positions would be included in the negative side of the circuit?
         [BEE(F), stud46]

In this situation T wants to check if S really understands the rule of observing polarity. To do this, T poses a question about positive and negative sides of the circuit. How should we model such a situation? First, we will assume that T's strategy of questioning may be represented by an e-scenario. If we do so, it is necessary to assume that T's beliefs and knowledge will be somehow reflected in the premises of the e-scenario. These beliefs and knowledge will concern the issues raised by tutorial dialogue context (namely understanding the problem under discussion). The premises relative to our problem might be expressed as

---

[4] See (Łupkowski, 2011a) and (Urbański and Łupkowski, 2010b).

sufficient conditions of understanding a tutorial situation. In a case such as this T would formulate his/her premises according to the following scheme:

'If S correctly solves task X, then S understands the problem under discussion'.

I will abbreviate this schema as $B \rightarrow A$ where A stands for 'S understands the problem under discussion', and B stands for 'S correctly solves task X'. The premises representing T's beliefs can be expressed by the following formulas: $B_1 \rightarrow A, \ldots, B_n \rightarrow A, \neg B_1 \wedge \ldots \wedge \neg B_n \rightarrow \neg A$. When we agree that T uses sufficient conditions of understanding the problem under discussion, then if S fails in all of them T becomes sure that S does not understand the problem. This seems to be a very intuitive interpretation especially when it comes to a tutorial situation.

The e-scenario built on the basis of premises of this kind is presented in Figure 3.3. One may notice that T will use the following procedure in this case: first T poses himself/herself a question of whether S understands the problem under discussion. Then T poses questions concerning conditions of 'understanding the problem'. Let us stress here that questions of the form $?\{A, \neg A, B_i\}$ are not queries of the e-scenario under consideration (they are necessary for the model), but they play an important role since on the one hand, they are e-implied and on the other hand, they imply queries. If S does not fulfil any of the conditions T becomes certain that the answer to the main question of the e-scenario is negative. What we observe in example (18) is a part of the execution of the presented questioning plan schema. The initial problem for T (i.e. $?A$ in the schema) is 'Does S understand the rule for observing polarity?'. One of the initial premises for Tutor would be 'If S correctly solves the task concerning which tab position would be included in the negative side of the circuit in a given situation, then S understands the rule of observing polarity.'

We may also consider the manner of the necessary conditions for formulating the conditions of understanding a tutorial situation. T's premises would be formulated according to the following schema:

'If S understands the problem under discussion,
then S correctly solves task X.'

By analogy to the former case, I will abbreviate this schema as $A \rightarrow C$, where A stands for 'S understands the problem under discussion', and C stands for 'S correctly solves task X'. The premises representing T's beliefs may be expressed with the following formulas: $A \rightarrow C_1, \ldots, A \rightarrow C_n, C_1 \wedge \ldots \wedge C_n \rightarrow A$. (where A is different from any of C_i ($1 \leq i \leq n$)).

One of the possible e-scenarios which might be used as a strategy for T in this case is presented in Figure 3.4. In this case T's strategy is to check the previously formulated conditions one by one. If S fulfils every single one, then T becomes sure that S understands the problem under discussion. If S fails in

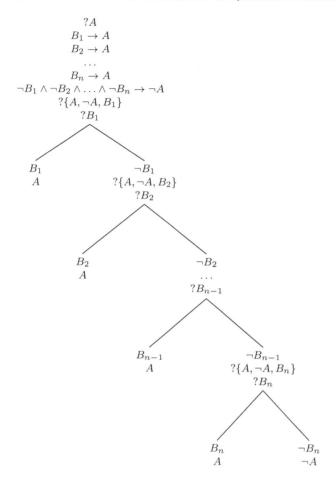

Fig. 3.3. Tutor's questioning plan for the sufficient conditions of understanding the problem under discussion

at least one of the conditions, he is assumed not to have understood the given problem. Let us consider example (19).

(19)   TUTOR: They are those red and black clips. You need to attach them to the circuit. Do you know where to attach them?

STUDENT: do you attach them to the positive and negative points?

TUTOR: No. Let's think about what it is that you are trying to accomplish by attaching them to the circuit. Do you know what it is that you are trying to accomplish?

STUDENT: I think that I am trying to somehow measure current by setting up the circuit.

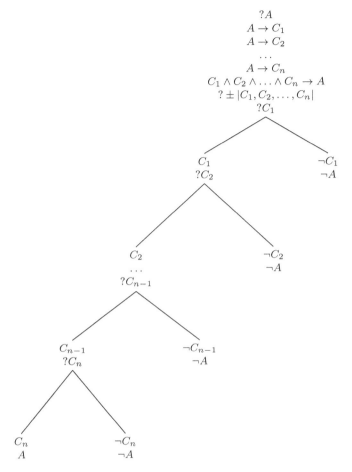

Fig. 3.4. Tutor's questioning plan for the necessary conditions of understanding the problem under discussion

> TUTOR: That's right. Now, in order to measure current, two things have to be true. The first thing is that current must be flowing. What needs to be the case in order for current to flow?
>
> STUDENT: am I to clip the red and black wires to wire 2?
>
> TUTOR: What do you believe would be accomplished by doing that?
>
> STUDENT: they would possibly complete the circuit.
>
> TUTOR: How would current flow in the circuit then? (what I mean is: where would it start? then where would it go? then where? etc.)

> STUDENT: maybe I should clip the red and black wires to area number 5 and 4?
>
> TUTOR: That is correct. Which position would you attach the red clip to?
>
> STUDENT: position number 4
>
> TUTOR: Good. OK, it seems you know how to hook the leads up...
>
> [BEE(F), stud47]

In this example we can recognise the strategy just described. T wants to check if S knows/understands how to attach clips to the circuit (20a). To check this, T uses set of premises concerning this problem, and then asks three auxiliary questions (which express certain tasks for S: (20b), (20c), (20d)). After obtaining satisfactory answers to these questions T concludes that S really understands the initial problem, which is clearly expressed by the end of example (20e).

(20) a. TUTOR: Do you know where to attach them? Let's think about what it is that you are trying to accomplish by attaching them to the circuit. [INITIAL PROBLEM]

   b. TUTOR: Do you know what it is that you are trying to accomplish? [AUX Q1]
      STUDENT: I think that I am trying to somehow measure current by setting up the circuit.
      TUTOR: That's right. [ANSWER ACCEPTANCE]

   c. TUTOR: What needs to be the case in order for current to flow? [AUX Q2]
      STUDENT: am I to clip the red and black wires to wire 2?
      TUTOR: What do you believe would be accomplished by doing that?
      STUDENT: they would possibly complete the circuit.
      TUTOR: How would current flow in the circuit then? (what I mean is: where would it start? then where would it go? then where? etc.)
      STUDENT: maybe I should clip the red and black wires to area number 5 and 4?
      TUTOR: That is correct. [ANSWER ACCEPTANCE]

   d. TUTOR: Which position would you attach the red clip to? [AUX Q3]
      STUDENT: position number 4
      TUTOR: Good. [ANSWER ACCEPTANCE]

   e. TUTOR: OK, it seems you know how to hook the leads up... [CONCLUSION]

Let us now get back to example (14) presented on page 56. This time we will employ the language $\mathcal{L}^?_{\mathbf{K}3}$ introduced in Chapter 2.3.2 to fully grasp the use of the "I do not know" statement in the dialogue.[5] Analysis of dialogues in educational contexts reveal that a teacher may use a kind of strategy of searching/provoking 'I don't know' type of answers to identify a student's lack of knowledge/understanding. After a situation such as this a teacher might ask a series of sub-questions that may lead the student to a better understanding of a given topic. Often such auxiliary questions are accompanied with some additional information about the topic.

As we can see, when T asks S the first question each option seems equally possible from the point of view of S. Thus the initial question may be reconstructed as follows:

$$?\{p \wedge q, p \wedge \neg q, p \wedge \boxminus q, \neg p \wedge q, \neg p \wedge \neg q, \neg p \wedge \boxminus q, \boxminus p \wedge q, \boxminus p \wedge \neg q, \boxminus p \wedge \boxminus q\}$$

where p stands for 'current flows through the wire' and q—'current flows through the meter'. Remember that this type of question is symbolised by $? \pm \boxminus |p, q|$. We can also identify T's premise explicated in the dialogue; it falls under the schema: $r \leftrightarrow \neg(p \wedge q)$, where r stands for 'the leads obstruct the current flow'. Now the dialogue may be modelled by the e-scenario presented in Figure 3.5.

The scenario relies on logical facts (3.6a), (3.6b), (3.6c). The vertical dots indicate that something more would happen in the e-scenario, if the Student's answers were different.

$$\mathsf{Im}(? \pm \boxminus |p, q|, ?\{p \wedge \neg q, \neg p \wedge q, p \wedge q, \neg p \wedge \neg q, \boxminus p \vee \boxminus q\})$$
(3.6a)

$$\mathsf{Im}(?\{p \wedge \neg q, \neg p \wedge q, p \wedge q, \neg p \wedge \neg q, \boxminus p \vee \boxminus q\}, ?\{p, \neg p, \boxminus p\})$$
(3.6b)

$$\mathsf{Im}(?\{p \wedge \neg q, \neg p \wedge q, p \wedge q, \neg p \wedge \neg q, \boxminus p \vee \boxminus q\}, \{r \leftrightarrow \neg(p \wedge q)\}, ?\{r, \neg r, \boxminus r\})$$
(3.6c)

After the Tutor asks the first question ('Can you tell me how current would flow?') he/she reformulates it, which is reflected in the e-scenario by the question $?\{p \wedge \neg q, \neg p \wedge q, p \wedge q, \neg p \wedge \neg q, \boxminus p \vee \boxminus q\}$ ('Would it flow only through the wire? Or only through the meter? Or through both? Or through neither?'). Now five answers are possible, and our Student chooses the last one ('I don't know' is represented by '$\boxminus p \vee \boxminus q$'). Then, to simplify the matter, the Tutor asks about p ('Is there any reason why it wouldn't flow through the wire?') again. After another Student's '$\boxminus p$' answer, the Tutor introduces an additional declarative premise $(r \leftrightarrow \neg(p \wedge q))$, which leads to new auxiliary question concerning r. This time S is able to provide an answer to the question about r (which is

---

[5] I adopt this analysis following (Leszczyńska-Jasion and Łupkowski, 2016).

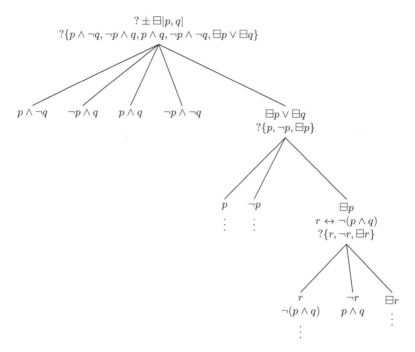

Fig. 3.5. E-scenario for the exemplary dialogue (14)

positive) and therefore by using the just introduced declarative premise S is able to provide the answer to the initial question, which is $p \wedge q$ ('The current would flow through both').

3.3. Summary

In this chapter natural language examples obtained from the BNC and BEE language corpora were analysed using IEL concepts (e-implication and e-scenario). The notion of e-implication appears to be well-defined and suited to the task of explaining the occurrence of dependent questions in dialogue. The notion serves as a rationale for certain dialogue moves. We show examples where certain parts of dialogue have to be reconstructed in order to grasp the move in question. We also analyse examples where all the necessary data are explicated. What is interesting is that we can put our analysis in the context of the teaching process (as especially suggested by the last example). This may lead us to the pragmatic interpretation of e-scenarios provided by IEL (and based on the notion of e-implication) used to analyse questioning strategies used in tutorial dialogues.

Chapter 4

Dialogic logic and dynamisation of e-scenarios execution

In this chapter I will be interested in information seeking dialogues, where one dialogue participant seeks the answer to some question(s) from other participant(s), who is believed to know the answer(s) (see McBurney and Parsons, 2002). Here we will focus on a particular kind of these dialogues, namely the ones where a problem is solved *via* the questioning process. An agent solves the problem by dividing it into several simpler sub-problems and by gathering solutions to these sub-problems from other agents, thus obtaining a solution to the initial problem (see the *Erotetic Decomposition Principle* described on page 27). I propose here a system that combines an IEL approach with a dialogue logic. For this purpose we will consider a dialogue as a certain game, which consists of a sequence of locution events (see Girle, 1997). As a result we obtain a modular system (which is easy to modify and adapt) that allows us to describe a game where one of the players is using e-scenarios as a questioning strategy.

The motivation for such an approach is twofold. Firstly, it comes from the system presented by Peliš and Majer (2010, 2011) (described also in Chapter 1.2.2), where a dynamic epistemic component is added to a logic of questions in order to describe an information seeking procedure. The key difference is that we will not use dynamic epistemic logic for this purpose, but dialogue logic inspired by Girle's DL systems (Girle, 1997). Analysing an interrogator's perspective in a dialogue in terms of commitment stores allows for a simpler and more intuitive description of the dialogue events. What is important is that we take into account the information expressed in a dialogue. In the presented framework we do not hypothesise about all the knowledge that is at the agents' disposal. This approach is much more convincing when we think of modelling natural language dialogues and also for systems of formal dialogues. Secondly, my approach is inspired by the system proposed in (Kacprzak and Budzynska, 2014; Budzynska et al., 2015). My approach shares the main intuition of using a logic in the background of a formal dialogue system in order to check the correctness of certain dialogue moves.

Let us remember that when we are using an e-scenario as a questioning strategy only queries of such an e-scenario are publicly asked by the agent who seeks the information.

As was mentioned before, an e-scenario represents a certain map of possible courses of events in a given questioning process. Each path of an e-scenario is one of the ways the process might go depending on the answers obtained to queries. One may imagine that executing such an e-scenario is simply eliminating these branches that are no longer possible after obtaining an answer to a given query. At the end of such an execution we will be left with only one path leading from the initial question through the auxiliary questions and answers to them to the solution to the initial question. Such a path will be called *the activated path of an e-scenario*.

The analysis presented in this chapter employs the language $\mathcal{L}^?_{\mathbf{K}3}$ and relies on concepts introduced in Chapter 2.3.2.

4.1. Two party games—the logic $DL(IEL)_2$

Let us imagine a dialogue between two participants: Bob and Alice. This dialogue is a questioning game in which Bob wants to solve a certain problem expressed by his initial question. He decomposes the initial question into series of auxiliary questions, that will be asked of Alice. By obtaining answers to auxiliary questions Bob hopes to solve the initial problem. We assume that he will use an e-scenario as his questioning strategy. We also assume that Alice is providing information according to the best of her knowledge. She may, however, withdraw from providing an answer.

Intuitively we may describe the game as follows. Bob asks questions and Alice provides answers. The game goes step-by-step. In one step Bob is allowed to:

– ask a question, or
– provide a justification for a question/solution, or
– end the game.

  Alice in one step can:

– provide an answer to a question;
– challenge the question ("I claim that your question is irrelevant", which in our case is understood as not fulfilling the conditions of e-implication);
– challenge the solution ("Please explain the solution you propose");
– deny answering ("I don't want to answer").

In a scenario where one of his questions is being attacked Bob is obliged to provide erotetic justification for the question (i.e. reveal a part of his questioning strategy).

We now propose a dialogue logic $DL(IEL)_2$ in order to precisely describe the rules of the game and the rules of executing Bob's questioning strategy. We will specify:

1. The taxonomy of locutions.
2. Commitment store rules.
3. Interaction rules.

A dialogue is a k-step finite game between Bob and Alice. Each move consists of a locutionary event performed by one of the players (done one-by-one). The first move is always performed by Bob, and it is a question.

Each step in a game will be presented as follows:

$$\langle n, \phi, \psi, U \rangle \tag{4.1}$$

where

n $(1 \leq n \leq k)$ is a number of the step of the dialogue;
ϕ is an agent producing the utterance;
ψ is an addressee;
U is the locution of the agent ϕ.

4.1.1. The taxonomy of locutions for $DL(IEL)_2$

The following types of locution are allowed in $DL(IEL)_2$:

Categorical statement A, $\neg A$, $A \wedge B$, $A \leftrightarrow B$, $A \leftrightarrow B$, and $\boxminus A$. These are responses given by Alice or solutions declared by Bob at the end of the game. As such they are d-wffs.

Question $?\{A_1, A_2, ..., A_n\}$. Questions asked by Bob. These are e-wffs.

Logical statement   Justifications provided by Bob: (i) stating that e-implication holds between a certain previous question (and possibly a set of declarative premises) and a question being challenged by Alice; and (ii) justifications for a challenged solution (in a form of an e-derivation for this solution).

Challenge   Alice's attack on a question asked by Bob, or on a solution provided by Bob at the end of a game. The intuition about this type of locution is that Alice claims that a question asked by Bob is irrelevant. In the case of a solution, Alice's challenge should be understood as a request for an explanation of it.

Withdrawal   Alice's statement, that she does not want to answer A. This statement is intuitively different than admitting a lack of knowledge with respect to A. Withdrawal is treated here as a non-cooperative locution. From the Bob's perspective however, for the two agent game, it will play the same role as an $\boxminus A$ answer.

Let us now introduce abbreviations for locutions described above in order to make descriptions simpler.

Bob's locutions	Alice's locutions
Q_i—question	$ANS(Q_i)$—categorical statement (answer to a question Q_i)
$LS(Q_i)$—logical statement for question Q_i	$WD(Q_i)$—withdrawal
SOL—solution to the initial problem (a categorical statement)	$CH(Q_i)$—challenge the question Q_i
$\boxminus SOL$—stating that the solution is not available	$CH(SOL)$—challenge the solution provided by Bob
$LS(SOL)$—logical statement for a solution	$ACC(SOL)$—accept the solution provided by Bob

4.1.2. Interaction rules for $DL(IEL)_2$

Interaction rules allow us to specify the agent's behaviour in our game.

(In1) (`Repstat`) No statement may occur if it is in the commitment store of both participants.

This rule prevents pointless repetitions.

(In2) (`Q-response`) When $\langle n, Bob, Alice, Q_i \rangle$, then

1. $\langle n+1, Alice, Bob, ANS(Q_i) \rangle$; or
2. $\langle n+1, Alice, Bob, CH(Q_i) \rangle$; or
3. $\langle n+1, Alice, Bob, WD(Q_i) \rangle$.

The rule states that when Bob asks a certain question, Alice may react by (1) simply providing a direct answer to this question; (2) by challenging the question (i.e. demanding an erotetic justification for the question); or (3) by withdrawal—i.e. informing Bob, that she does not want to answer the question.

(In3) (`Q-challenge`) When $\langle n, Alice, Bob, CH(Q_i) \rangle$, then

1. $\langle n+1, Bob, Alice, LS(Q_i) \rangle$.

The rule regulates Bob's reaction to Alice challenging the question asked. He is obliged to provide a logical statement for that question (in our case it should be a statement of the form: $\mathsf{Im}(Q, X, Q_i)$, where Q_i is the challenged question). The intuition is that Bob will state that the challenged question is e-implied by one of the previous questions (possibly on the basis of Bob's declarative premises and answers already provided by Alice).

(In4)   (Q-ChallengeResp) When $\langle n, Bob, Alice, LS(Q_i) \rangle$, then

1. $\langle n+1, Alice, Bob, ANS(Q_i) \rangle$; or
2. $\langle n+1, Alice, Bob, WD(Q_i) \rangle$.

After the logical statement for the challenged question is provided by Bob, Alice may (1) provide the answer or (2) withdraw. We assume that both Alice and Bob accept the normative yardstick for erotetic reasoning provided by e-implication. Because of this, we may say that Alice provided with the logical statement for the challenged question of the form proposed in (In3) will accept the relevance of the justified question.

(In5)   (SOLReaction) When $\langle n, Bob, Alice, SOL \rangle$, then

1. $\langle n+1, Alice, Bob, ACC(SOL) \rangle$ and game ends;
2. $\langle n+1, Alice, Bob, CH(SOL) \rangle$.

This is the regulation of Alice's reaction to a solution being declared by Bob. Alice may (1) ask for a justification (challenge the solution) or (2) state that she agrees/accepts it—which will end the instance of the game.

(In6)   (SOL-ChallengeResp) When $\langle n, Alice, Bob, CH(SOL) \rangle$, then

1. $\langle n+1, Bob, Alice, LS(SOL) \rangle$ and game ends $(n+1 = r)$.

SOL-ChallengeResp regulates justification of the solution move for Bob. Analogically to Q-ChallengeResp, Bob should provide a logical statement for the declared solution. In our case, such a logical statement is the activated path of an e-scenario used by Bob in the game.

(In7)   (IgnoranceResp) When $\langle n, Alice, Bob, ans(Q_i) = \boxminus A \rangle$, then

1.   Bob checks his strategy whether there is a successful path of his e-scenario, if yes then $\langle n+1, Bob, Alice, Q_{i+1} \rangle$;
2.   if not, then $\langle n+1, Bob, Alice, \boxminus SOL \rangle$ and the game ends.

The rule says what to do in a scenario, where the answer provided by Alice is of the form $\boxminus A$, i.e. 'I have no knowledge with respect to A.' To take his move, Bob first checks his e-scenario for the successful path (i.e. is there a path leading from the point where he is in the current state of the game, leading to a leaf of the e-scenario, which is not marked with $\boxminus$). If such a path exists (Bob can reach the solution to the initial question despite the gaps in the information provided by Alice), then (1) Bob asks the next auxiliary question from that path of his e-scenario. If not, (2) Bob declares that he does not know the solution and the instance of the game ends.

(In8)   (NoSol) When $\langle n, Alice, Bob, WD(Q_i) \rangle$, then

1.   Bob checks his strategy whether there is a successful path of his e-scenario, if yes then $\langle n+1, Bob, Alice, Q_{i+1} \rangle$;
2.   if not, then $\langle n+1, Bob, Alice, \boxminus SOL \rangle$ and the game ends.

In the two party game where no information sources except Alice are available for Bob the routine of actions for NoSol will be the same as for

IgnoranceResp. After Alice withdraws stating that she does not want to answer the question Q_i, Bob checks his e-scenario for a successful path. If such a path exists, Bob asks the next auxiliary question from that path. If there is no such path, then Bob declares that he does not know the solution and ends that instance of the game. The reason why the distinction between **(In7)** and **(In8)** is preserved here, is that intuitions behind NoSol and IgnoranceResp are different at the pragmatic level. What is more, this distinction will play an important role in multi-party dialogues.

4.1.3. Commitment Store Rules for $DL(IEL)_2$

Let us now have a look at the commitment store (ComSt) rules for $DL(IEL)_2$. They state which locutions are added $(+)$ or subtracted $(-)$ from players' ComSt during the game.

Rule	Locution	Bob's ComSt	Alice's ComSt
(CS1)	Q_i	$+Q_i$	$+Q_i$
(CS2)	$ANS(Q_i)$	$+ANS(Q_i)$	$+ANS(Q_i)$
(CS3)	$CH(Q_i)$	$-Q_i$ $+CH(Q_i)$	$-Q_i$ $+CH(Q_i)$
(CS4)	$LS(Q_i)$	$+Q_i$ $-CH(Q_i)$	$+Q_i$ $+LS(Q_i)$
(CS5)	$WD(Q_i)$	$+\boxminus ans(Q_i)$	$+WD(Q_i)$
(CS6)	SOL	$+SOL$	$+SOL$
(CS7)	$CH(SOL)$	$-SOL$ $+CH(SOL)$	$-SOL$ $+CH(SOL)$
(CS8)	$LS(SOL)$	$+SOL$ $-CH(SOL)$	$+SOL$ $+LS(SOL)$

The intuitions behind these rules are the following:

(CS1) When a question is asked by Bob it goes into both players' ComSt.
(CS2) When an answer is provided by Alice it goes into ComSt of both players.
(CS3) If the question asked by Bob is challenged, it should be removed from both players' ComSt (this will allow us to ask it again in further moves). The information about the challenge is added.
(CS4) According to **(In4)**, Bob should react to a challenge to a question by providing a logical statement for this question. In terms of his ComSt the information about the challenge is removed $(-CH(Q_i))$. As for Alice,

she adds the logical statement for Q_i and Q_i itself to her ComSt. There is no need to add $LS(Q_i)$ to Bob's ComSt, because this logical statement is generated on the basis of the activated part of an e-scenario used by Bob. Notice that $CH(Q_i)$ is not removed from Alice's ComSt (this prevents Alice from challenging one question more than one time—see the `RepStat` interaction rule).

(CS5)   When Alice does not want to provide an answer to Q_i ($WD(Q_i)$), then the information about a lack of knowledge when it comes to answer $ans(Q_i)$ is added to Bob's ComSt (this allows Bob to take actions according to the `NoSol` interaction rule), and the information about the withdrawal goes to Alice's ComSt.

(CS6)–(CS8)   These rules concerning solutions are analogous to **(CS2)**, **(CS3)** and **(CS4)**.

In terms of e-scenario execution we may say that the ComSt rules indicate Bob's moves in the game. We may imagine an e-scenario as a game-board on which Bob performs moves accordingly to the content of his ComSt.

4.1.4. $DL(IEL)_2$ Example

Let us now consider an example of a game described in $DL(IEL)_2$. We will use the following tabular form of reporting steps of a game:

Move	Bob's ComSt	Alice's ComSt	Interaction rule
$\langle n, \phi, \psi, U \rangle$	$+/-U$	$+/-U$	In1–In8 (if any)

Let us assume that in our example Bob's questioning strategy is represented by the following e-scenario:

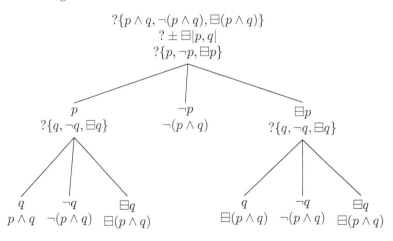

In building the e-scenario the following facts were used:

$$\mathsf{Im}(?\{p \land q, \neg(p \land q), \boxminus(p \land q)\}, ? \pm \boxminus|p, q|) \qquad (4.2\text{a})$$

$$\mathsf{Im}(? \pm \boxminus|p, q|, ?\{p, \neg p, \boxminus p\}) \qquad (4.2\text{b})$$

$$\mathsf{Im}(? \pm \boxminus|p, q|, ?\{q, \neg q, \boxminus q\}) \qquad (4.2\text{c})$$

In each step Bob uses the e-scenario to ask the next auxiliary question. To make a move Bob checks his ComSt for the last categorical statement and asks the next auxiliary question downward the chosen path of the e-scenario.

We will now exemplify this set up on the basis of the following games.

4.1.4.1. Game 1: simple case

Move	Bob's ComSt	Alice's ComSt	Int rule
$\langle 1, Bob, Alice, ?\{p, \neg p, \boxminus p\}\rangle$	$+?\{p, \neg p, \boxminus p\}$	$+?\{p, \neg p, \boxminus p\}$	—
$\langle 2, Alice, Bob, p\rangle$	$+p$	$+p$	In2.1
$\langle 3, Bob, Alice, ?\{q, \neg q, \boxminus q\}\rangle$	$+?\{q, \neg q, \boxminus q\}$	$+?\{q, \neg q, \boxminus q\}$	—
$\langle 4, Alice, Bob, q\rangle$	$+q$	$+q$	In2.1
$\langle 5, Bob, Alice, SOL = p \land q\rangle$	$+p \land q$	$+p \land q$	—
$\langle 6, Alice, Bob, ACC(SOL)\rangle$	—	—	In5.1

This game runs without any complications. In the first step, Bob asks the first query from his e-scenario, namely $?\{p, \neg p, \boxminus p\}$. Alice provides answer p, which is added to Bob's ComSt. This allows him to go down the e-scenario and ask the next query: $?\{q, \neg q, \boxminus q\}$. Alice responses with q. At this point of the game Bob has all the information needed to answer his initial question. All that is left to do is to announce the solution, which is done in the 5th step of the game. In the last step, Alice accepts the solution provided by Bob in the previous step.

The game activates the leftmost path of the e-scenario used by Bob which is depicted in the following diagram:

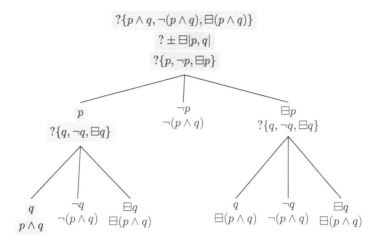

4.1.4.2. Game 2: I do not know

In this game Bob will encounter an 'I do not know' answer.

Move	Bob's ComSt	Alice's ComSt	Int rule
$\langle 1, Bob, Alice?\{p, \neg p, \boxminus p\}\rangle$	$+?\{p, \neg p, \boxminus p\}$	$+?\{p, \neg p, \boxminus p\}$	—
$\langle 2, Alice, Bob, \boxminus p\rangle$	$+\boxminus p$	$+\boxminus p$	In2.1
$\langle 3, Bob, Alice, ?\{q, \neg q, \boxminus q\}\rangle$	$+?\{q, \neg q, \boxminus q\}$	$+?\{q, \neg q, \boxminus q\}$	In7.1
$\langle 4, Alice, Bob, \neg q\rangle$	$+\neg q$	$+\neg q$	In2.1
$\langle 5, Bob, Alice, SOL = \neg(p \wedge q)\rangle$	$+\neg(p \wedge q)$	$+\neg(p \wedge q)$	—
$\langle 6, Alice, Bob, ACC(SOL)\rangle$	—	—	In5.1

In the 3-rd step the rule (**In7 IgnoranceResp**) is used: Bob checks his strategy (whether there is a successful path of his e-scenario, if yes then $\langle n + 1, Bob, Q_i\rangle$—i.e in his next move Bob asks the next auxiliary question from his e-scenario). In Bob's scenario there is one successful path (leading to a solution): $?\{q, \neg q, \boxminus q\}$; $\neg q$; $\neg(p \wedge q)$, so he can act according to (**In7.1**), i.e. ask the next auxiliary question.

The e-scenario path activated by this game is presented in the diagram below.

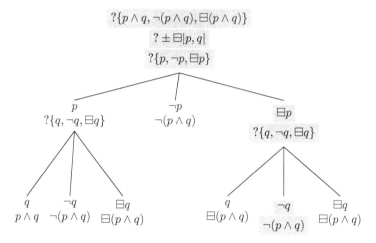

4.1.4.3. Game 3: question challenge

Now let us consider the case in which Alice decides to challenge a question.

Move	Bob's ComSt	Alice's ComSt	Int rule
$\langle 1, Bob, Alice,$ $?\{p, \neg p, \boxminus p\}\rangle$	$+?\{p, \neg p, \boxminus p\}$	$+?\{p, \neg p, \boxminus p\}$	—
$\langle 2, Alice, Bob, \boxminus p\rangle$	$+\boxminus p$	$+\boxminus p$	In2.1
$\langle 3, Bob, Alice,$ $?\{q, \neg q, \boxminus q\}\rangle$	$+?\{q, \neg q, \boxminus q\}$	$+?\{q, \neg q, \boxminus q\}$	In7.1
$\langle 4, Alice, Bob,$ $CH(+?\{q, \neg q, \boxminus q\})\rangle$	$-?\{q, \neg q, \boxminus q\}$ $+CH(+?\{q, \neg q, \boxminus q\})$	$-?\{q, \neg q, \boxminus q\}$ $+CH(+?\{q, \neg q, \boxminus q\})$	In2.2
$\langle 5, Bob, Alice,$ $LS(?\{q, \neg q, \boxminus q\})\rangle$	$+?\{q, \neg q, \boxminus q\}$ $-CH(+?\{q, \neg q, \boxminus q\})$	$+?\{q, \neg q, \boxminus q\}$ $-CH(+?\{q, \neg q, \boxminus q\})$	In3
$\langle 6, Alice, Bob, \neg q\rangle$	$+\neg q$	$+\neg q$	In4.1
$\langle 7, Bob, Alice,$ $SOL = \neg(p \wedge q)\rangle$	$+\neg(p \wedge q)$	$+\neg(p \wedge q)$	—
$\langle 8, Alice, Bob,$ $ACC(SOL)\rangle$	—	—	In5.1

Where the $LS(?\{q, \neg q, \boxminus q\})$ is of the following form:

$$LS(?\{q, \neg q, \boxminus q\}) = \mathsf{Im}(? \pm \boxminus | p, q |, \boxminus p, ?\{q, \neg q, \boxminus q\})\rangle.$$

In the 4th step Alice challenges the question $?\{q, \neg q, \boxminus q\}$. According to the rule (**In3**) Bob is obliged to provide a justification for this question, i.e. e-implication that holds, for this question, for a given situation. This is done in step 5. After LS for the challenged question is given, Alice provides an answer to the question (**In4.1**; the other option for her would be to withdraw—**In4.2**).

As a result of the game the same path of the e-scenario gets activated as in the case of Game 2.

4.1.4.4. Game 4: solution challenge

Our last game considers a situation where Alice challenges the solution.

Move	Bob's ComSt	Alice's ComSt	Int rule
$\langle 1, Bob, Alice, ?\{p, \neg p, \boxminus p\}\rangle$	$+?\{p, \neg p, \boxminus p\}$	$+?\{p, \neg p, \boxminus p\}$	—
$\langle 2, Alice, Bob, \neg p \rangle$	$+\neg p$	$+\neg p$	In2.1
$\langle 7, Bob, Alice, SOL = \neg(p \wedge q)\rangle$	$+\neg(p \wedge q)$	$+\neg(p \wedge q)$	—
$\langle 8, Alice, Bob, CH(\neg(p \wedge q))\rangle$	$- \neg(p \wedge q)$ $+CH(\neg(p \wedge q))$	$- \neg(p \wedge q)$ $+CH(\neg(p \wedge q))$	In5.2
$\langle 9, Bob, Alice, LS(\neg(p \wedge q))\rangle$	$+\neg(p \wedge q)$ $-CH(\neg(p \wedge q))$	$+\neg(p \wedge q)$ $+LS(\neg(p \wedge q))$	In6

Where the LS is obtained from the activated path of the e-scenario used by Bob and it is of the following form:

$$LS(\neg(p \wedge q)) = ?\{p \wedge q, \neg(p \wedge q), \boxminus(p \wedge q)\}; \ ? \pm \boxminus | p, q |; \ ?\{p, \neg p, \boxminus p\}; \ \neg p; \ \neg(p \wedge q).$$

The activated part of the e-scenario is marked on the schema below.

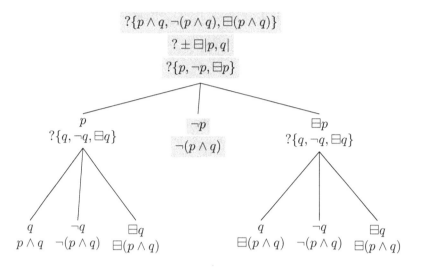

4.2. Multi-agent games—the logic $DL(IEL)_{mult}$

In our case a multi-agent game is a situation, where Bob employs more than one information source (IS) in his questioning process.

One of the most natural contexts for such a situation to appear is when an IS withdraws from providing an answer to a question $(WD(Q_i))$. Bob's reaction would be to start a new sub-game with another player in order to retrieve the missing data. We may imagine two strategies of executing such a sub-game:

1. Bob's starts a sub-game to ask one question to another player (i.e. the question that Ann withdrew from answering) and then comes back immediately into the main game.
2. Bob starts a sub-game and continues it until reaching:

   a. a solution to his main problem;
   b. a withdrawal from the new player.

In each case we need to introduce an indexing of formulas from Bob's ComSt. Index will point out the information source for each piece of information gathered during the game. This also allows us to say that the solution obtained after executing the e-scenario will be *relative* to the information sources.

4.2.1. $DL(IEL)_{mult}$

A dialogue is a k-step finite game between Bob and a finite number of agents (information sources) $IS_1, ..., IS_n$. Each move is done by a locution event performed by one of the players (done one-by-one). We assume that there is no communication between information sources. It is Bob who asks questions and information sources are providing answers. Bob always asks only one information source at a time.

4.2.1.1. $DL(IEL)_{mult}$ locutions

The following types of locution are allowed in $DL(IEL)_{mult}$ (they are analogous as locutions allowed for $DL(IEL)_2$—see page 69):

Categorical statement A, $\neg A$, $A \wedge B$, $A \leftrightarrow B$, $A \leftrightarrow B$, and $\boxminus A$. These are responses given by IS.

Question $?\{A_1, A_2, ..., A_n\}$. Questions asked by Bob.

Logical statement   Justifications provided by Bob, stating that an e-implication holds between certain question and question being attacked by an agent who is an IS.

Challenge   An IS attack on a question asked by Bob.

Withdrawal   An IS statement, that she does not want to answer a question.

Let us remember that each move in the game is presented as follows:

$$\langle n, \phi, \psi, U \rangle$$

where

n ($1 \leq n \leq k$) is a number of the step of the dialogue;
ϕ is an agent producing the utterance;
ψ is an addressee;
U is the locution of the agent ϕ.

4.2.1.2. $DL(IEL)_{mult}$ interaction rules

Interaction rules 1–4 are the same as for two-party games (see Section 4.1.2). The changes are introduced for the rules 5–8 in order to manage the game with many information sources.

(In1)   (`Repstat`) No statement may occur if it is in the commitment store of any of the participants.

(In2)   (`Q-response`) When $\langle n, Bob, IS_i, Q_i \rangle$, then

1. $\langle n+1, \mathsf{IS}_i, Bob, ans(Q_i)\rangle$; or
2. $\langle n+1, \mathsf{IS}_i, Bob, CH(Q_i)\rangle$; or
3. $\langle n+1, \mathsf{IS}_i, Bob, WD(Q_i)\rangle$.

(In3) (Q-challenge) When $\langle n, \mathsf{IS}_i, Bob, CH(Q_i)\rangle$, then

1. $\langle n+1, Bob, \mathsf{IS}_i, LS(Q_i)\rangle$.

(In4) (Q-ChallengeResp) When $\langle n, Bob, \mathsf{IS}_i, LS(Q_i)\rangle$, then

1. $\langle n+1, \mathsf{IS}_i, Bob, ans(Q_i)\rangle$; or
2. $\langle n+1, \mathsf{IS}_i, Bob, WD(Q_i)\rangle$.

(In5) (IgnoranceResp) When $\langle n, \mathsf{IS}_i, Bob, ans(Q_i) = \boxdot A\rangle$, then

1. Bob checks his strategy (whether there is a successful path of his e-scenario, if yes then $\langle n+1, Bob, \mathsf{IS}_i, Q_{i+1}\rangle$;
2. if not, then $\langle n+1, Bob, \mathsf{IS}_{i+1}, ?Q_i\rangle$; Bob starts a sub-game with another information source.

(In6) (NoSol) When $\langle n, \mathsf{IS}_i, Bob, WD(Q_i)\rangle$, then

1. $\langle n+1, Bob, \mathsf{IS}_{i+1}, ?Q_i\rangle$; Bob starts a sub-game with another information source.

(In7) (SubGameEnd) rule to end a sub-game. Ending a sub-game depends on the purpose for which the sub-game was initiated.

1. If a sub-game is a result of (IgnoranceResp), then after obtaining a response to a given question, Bob ends the instance of a sub-game with a new information source and gets back to the main play with the first agent.
2. If a sub-game is a result of (NoSol), then Bob continues the sub-game with a new information source.

(In8) (GameEnd) The game ends when Bob announces the solution to all information sources involved in a dialogue.

4.2.1.3. $DL(IEL)_{mult}$ Commitment Store Rules

In the case of $DL(IEL)_{mult}$ we assume that Bob is communicating with information sources in a private way. This means that the exchange of information is not public—as a consequence only ComSts of players engaged in a certain part of a dialogue will change respectively. In Bob's ComSt we add indices to utterances in order to preserve the information about the information source.

Rule	Locution	Bob's ComSt	IS_i ComSt
(CS1)	Q_i	$+Q_i^{IS_i}$	$+Q_i$
(CS2)	$d(Q_i)$	$+ANS(Q_i)^{IS_i}$	$+ANS(Q_i)$
(CS3)	$CH(Q_i)$	$-Q_i^{IS_i}$ $+CH(Q_i)^{IS_i}$	$-Q_i$ $+CH(Q_i)$
(CS4)	$LS(Q_i)$	$+Q_i^{IS_i}$ $-CH(Q_i)^{IS_i}$	$+Q_i$ $+LS(Q_i)$
(CS5)	$WD(Q_i)$	$+WD(Q_i)^{IS_i}$	$+WD(Q_i)$
(CS6)	SOL	$+SOL^{IS_i}$	$+SOL$

4.2.2. $DL(IEL)_{mult}$ Example

Bob's questioning strategy is represented by the same e-scenario, as for the two agent game (see page 73).

Bob starts the game with the first information source IS_1. He asks the first query from his e-scenario, namely $?\{p, \neg p, \boxminus p\}$.

Move	Bob's ComSt	IS_1 ComSt	Int rule
$\langle 1, Bob, IS_1 ?\{p, \neg p, \boxminus p\}\rangle$	$+?\{p, \neg p, \boxminus p\}^{IS_1}$	$+?\{p, \neg p, \boxminus p\}$	—
$\langle 2, IS_1, WD(?\{p, \neg p, \boxminus p\})\rangle$	$+WD(?\{p, \neg p, \boxminus p\})^{IS_1}$	$+WD(?\{p, \neg p, \boxminus p\})$	In6

In the second step of the game IS_1 withdraws from answering Bob's question. As a consequence Bob starts a sub-game with $\mathbf{IS_2}$ according to rule (**In6**).

Move	Bob's ComSt	IS_2 ComSt	Int rule
$\langle 3, Bob, IS_2, ?\{p, \neg p, \boxminus p\}\rangle$	$+?\{p, \neg p, \boxminus p\}^{IS_2}$	$+?\{p, \neg p, \boxminus p\}$	In6
$\langle 4, IS_2, Bob, p\rangle$	$+p^{IS_2}$	$+p$	In2.1
$\langle 5, Bob, IS_2, ?\{q, \neg q, \boxminus q\}\rangle$	$+?\{q, \neg q, \boxminus q\}^{IS_2}$	$+?\{q, \neg q, \boxminus q\}$	In2.1
$\langle 6, IS_2, Bob, \boxminus q\rangle$	$+\boxminus q^{IS_2}$	$+\boxminus q$	In5.2

In this sub-game with IS_2 Bob gathers information about p in the 4th step. Unfortunately in the 6th step IS_2 declares the lack of knowledge with respect to q and at this point Bob has no successful path in his e-scenario (there is no path leading to the solution to his initial problem). That is why he needs to start a new sub-game with another information source. Observe that Bob will not get back to the initial source because the reason for quitting the game with IS_1 was a withdrawal (see `SubGameEnd` (**In7.1**)). What is more—according to the same rule (**In7.2**) Bob would get back to the IS_2, because the reason for leaving this game is only a lack of necessary information to solve the problem.

Bob starts another sub-game with IS_3.

Move	Bob's ComSt	IS_3 ComSt	Int rule
$\langle 7, Bob, IS_3, ?\{q, \neg q, \boxminus q\}\rangle$	$+?\{q, \neg q, \boxminus q\}^{IS_3}$	$+?\{q, \neg q, \boxminus q\}$	In5.2
$\langle 8, IS_3, Bob, q\rangle$	$+q^{IS_3}$	$+q$	In2

All the necessary information is collected at this point of the game, and Bob may reach the solution in his e-scenario. That is the reason why there is no need to come back to the sub-game with IS_2. The last thing to do is to announce the solution to all the agents involved in the information gaining—see rule (**In8**).

Move	Bob's ComSt	$IS_{\{1-3\}}$ ComSt	Int rule
$\langle 9, Bob, (IS_1, IS_2, IS_3, SOL = p \wedge q\rangle$	$+(p \wedge q)^{IS_1, IS_2, IS_3}$	$+p \wedge q$	In8

The obtained solution is relative to the following information sources: IS_1, IS_2, IS_3. One may observe that a simple accounting of all agents involved in a dialogue with Bob (i.e. take all the indices that are present in Bob's ComSt) results in also counting IS_1. This might not be an intuitive solution, because this agent actually withdrew from providing the answer to Bob's question. In order to avoid a situation like this we may only take into account indices of the declarative formulas from Bob's ComSt. This solution will exclude agents that did not contribute to the solution of the initial problem—see ComSt rules for $DL(IEL)_{mult}$. If we analyse the final declarative part of the Bob's ComSt we have:

Declarative formulas	Step in the game
p^{IS_2}	$\langle 4, IS_2, Bob, p\rangle$
$\boxminus q^{IS_2}$	$\langle 6, IS_2, Bob, \boxminus q\rangle$
q^{IS_3}	$\langle 8, IS_3, Bob, q\rangle$

In this case the solution is relative only to agents, who contributed important facts to solving the initial problem—$(p \land q)^{\mathsf{IS}_2, \mathsf{IS}_3}$.

4.3. Summary

In this chapter we introduced a general framework for modelling multi-agent information seeking dialogues with erotetic search scenarios. The framework is based on dialogue logic with underlying IEL concepts of erotetic inferences validity.

The framework is fairly universal and modular. It can be modified on the level of rules used to describe the dynamics of questioning process. E.g. we may add new locution types, or modify interaction rules to obtain the desired behaviours of our agents. What is more we can also adapt IEL tools used here in order to tailor them better to our needs (one can e.g. use the e-scenario based on the notion of weak e-implication (Urbański et al., 2016b), falsificationist e-implication (Grobler, 2012) or epistemic e-implication (Peliš and Majer, 2010, 2011), (Švarný et al., 2013)).

Chapter 5

Cooperation and questioning

In this chapter I present how a pragmatic interpretation of e-scenarios may be used with regard to cooperative answering in the field of databases and information systems. First, I will describe the idea of cooperative answering and the most popular techniques used in this area. After that I introduce techniques of generating cooperative answers on the basis of e-scenarios. These techniques allow for supplementing direct answers to the initial questions obtained after e-scenarios execution against a database with additional explanations useful for users. The next section of this chapter contains a proposal of enrichment regarding cooperative answering with cooperative questioning techniques (which also uses e-scenarios as its main tool). The idea is that a system would respond to a user's question with a question. The chapter ends with an introduction in dialogue logic implementing the presented techniques.

5.1. What is cooperative answering?

The issue of cooperative answering is important in the field of databases and information systems. It offers the possibility of more natural and effective interactions with these computer systems. This improves the user experience aspect. Databases and information systems in general offer correct answers to user's questions (as far as these systems contain valid data). The problem is to ensure that the answers will also be non-misleading and useful for a user. Generally, we can say that the systems in question are capable of providing direct answers (in the sense proposed by Webber 1985), i.e. information directly requested by the user's question. However, we would expect a cooperative response — which is understood as an informative reaction to the user's question — from such a system (cf. Webber 1985 and Gal 1988). On the basis of Grice's maxims (1975) three features of this type of cooperative response are pointed out (cf. Gaasterland et al. 1992, p. 3). This type of response should be:

(i)   correct,
(ii)   non-misleading, and
(iii) useful

answer to a query.

After (Gal, 1988, p. 2), let us consider a well-known example, which shall shed some light on what counts as a cooperative answer. Imagine that a student wants to evaluate a course before enrolling in it. He asks the following question:

Q_1:   How many students failed course number CS400 last semester?

Assume also that the course CS400 was not given last semester. For most database interface systems the answer to the student's question would be:

A_1^*:   None.

This answer is correct according to the database information state. But on the other hand, it is easy to notice that it is also misleading for the student (who is not aware of the fact that the course was not given in the last semester) and thus uncooperative from our perspective.

We may easily imagine how such an answer would lead to so-called *stonewalling* (see Godfrey, 1997). Our student, convinced that the course CS400 was not difficult might then ask

Q_2:   How many students passed course number CS400 last semester?

and the answer would be:

A_2^*:   None.

And this might be continued further. Our confused student might ask:

Q_3:   Who taught CS400 last semester?

and again the answer would be:

A_3^*:   No one.

However, when we think about a secretary answering the first question we may imagine that the secretary would easily recognise the student's false assumption and correct it in her answer:

A_1:   No one, but the reason for this is that course number CS400 was
         not offered last semester.

The answer A_1 is not only correct, but it is also non-misleading and useful for the student. The cooperative answer given by the secretary facilitates the student's further search. We may, for example, imagine that the next question— asked on the basis of A_1—would be: *When was CS400 offered for the last time and how many students failed it then?*

A similar idea is visible in the following examples (Gal, 1988, p. 142-143). Cooperative parts of responses are emphasised:

Q: Does Smith teach Claire?

A: Yes;

by the way, Smith teaches only in the English-as foreign-language department.

Q: Does professor Smith teach in the History department?

A: No.

Smith teaches only in the English-as-foreign-language department.

Yet another examples might be the ones actually provided by WEBCOOP (a cooperative question-answering system on the web) created by Benamara and Saint-Dizier (2003, p. 3). Imagine that a user wants to rent a country cottage. The only problem is that the user wants to rent this cottage in the Midi Pyrénées region and the cottage should be located by the seashore:

Q: Can I rent a country cottage in the Midi Pyrénées region by the seashore?

For most database interface systems the answer would be:

A: No.

And again, without additional explanations given, this response would be misleading for the user, and thus uncooperative from our perspective. We would expect an answer similar to:

A: No. *The Midi Pyrénées region is not by the seashore. I can offer you a country cottage in another region in France by the seashore, or in the Midi Pyrénées region.*

A similar schema of answering is visible in the following example:

Q: [Can I rent] A chalet in Corsica for 15 persons?

A: A chalet capacity is less than 10 persons in Corsica
Flexible solutions to go further:
1. *2 close-by chalets in Corsica*
2. *Another accommodation type: hotel, pension in Corsica*
3. *A 15-person country cottage in another region in France*

To solve the problem of providing cooperative answers, certain specific techniques were developed. The most important are the following:

– consideration of specific information about a user's state of mind,
– evaluation of presuppositions of a query,
– detection and correction of misconceptions in a query (other than a false presupposition),
– formulation of intensional answers,
– generalization of queries and of responses.

A detailed description of the above techniques may be found in (Gaasterland et al., 1992) and (Godfrey, 1997). For their implementation in various database and information systems see e.g. (Godfrey et al., 1994), (Gal, 1988), (Benamara and Saint Dizier, 2003).

In this chapter we will consider the following important aspect of cooperative answering: *providing additional information useful for a user confronted with a failure.* As Terry Gaasterland puts it:

> On asking a query, one assumes there are answers to the query too.
> If there are no answers, this deserves an explanation. (Gaasterland
> et al., 1992, p. 14)

I will address this issue, focusing my attention on cases where the answer to a question is negative, and there is no answer available in a database.

5.2. E-scenarios in generating cooperative answers—the $COOP_{IEL}$ system

In this section, I will introduce techniques that allow for the generation of co-operative answers of the type presented above, i.e. answers to users' queries enriched with additional explanations. For this purpose, I will assume that our exemplary database system is enriched with a cooperative layer. The architecture of such a system and its tasks are described in details below.

In the presented approach e-scenarios are applied by a layer located between a user and the DB (let us call it a 'cooperative layer'). E-scenarios are stored and processed in this layer. The cooperative layer proceeds a question asked by a user by carrying out the relevant auxiliary questions/queries against the DB in a way determined by an e-scenario. The received answers to queries are then transformed into an answer to the main question within the cooperative layer. The scheme of such a system is presented in Figure 5.1.

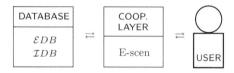

Fig. 5.1. Scheme of the cooperative database system using e-scenarios

The cooperative layer is responsible for:

– e-scenario generation,
– e-scenario storage,
– question analysis,
– e-scenario execution against a database.

The e-scenarios which I will use here exhibit certain properties characterised in details in Chapter 2.3.1.2. These will be e-scenarios relative to non-empty sets of initial premises (i.e. not pure e-scenarios). They will also be *complete* (each direct answer to the initial question will be the last element of a path of the e-scenario) and formulated in the *canonical form*, which means that all the declarative premises in these e-scenarios occur before the first query. What is most important, they will be *information-picking* e-scenarios. This means that the principal question of an e-scenario like this cannot be resolved by means of the declarative initial premises only and that no direct answer to a query is the answer to the initial question.

What I would like to propose here is a system that allows for interactions with a type of information system (which resembles the architecture of deductive databases). The system proposed here allows for generating cooperative answers by means of e-scenarios.

The cooperative layer copes with interactions between a user and a database in the following way:

1. First a user's question is analysed in the layer.
2. The cooperative layer interacts with the database.
3. After the interaction it analyses the data obtained and it may supplement the data with additional explanations useful for the user.

Let us firstly describe the architecture of our information system. It consists of:

- An extensional database ($\mathcal{EDB}$)—built out of facts,
- An intensional database ($\mathcal{IDB}$)—built out of rules.

Let us observe that the $\mathcal{IDB}$ part of the system allows for introducing and defining new concepts based on the facts present in the $\mathcal{EDB}$.

We impose the following restrictions on the $\mathcal{IDB}$. It may consist only of the rules of the form:

$$A \leftrightarrow B_1 \otimes ... \otimes B_n$$

where $\otimes$ is any of the connectives: $\wedge$, $\vee$, $\rightarrow$, $\leftrightarrow$; A is different than any of $B_1, ..., B_n$; and $B_1, ..., B_n$ are elements of $\mathcal{EDB}$.

The intuition behind such rules is that they allow for the introduction and defining of new concepts out of the facts from the $\mathcal{EDB}$. That is why the left side of such a rule cannot include any formula from the $\mathcal{EDB}$, while the right side of the rule is built out of facts only.

Let us consider a simple database with rules of this kind—I will refer to this database as DB.[1] It is presented in Table 5.1. DB contains facts about

---

[1] One may observe that although we have used predicates here, the whole reasoning may be perfectly modelled in the language of Classical Propositional Logic. We use this notation mainly for its readability.

a group of students. It states that a given person is a student $(st(x))$, and is signed up for classes $(sig(x, class))$ where the possible options are: philosophy courses ($phil1$, $phil2$) and linguistics courses ($ling$ and $cogling$). The rules of the DB introduce new concepts, namely they regulate when a student is:

- of a philosophical specialisation $(phspec(x))$—iff this student is signed up for both philosophy courses (rule 1);
- of linguistic specialisation $(lspec(x))$—iff the student is signed up for both linguistic courses (rule 2);
- advanced student $(advancedst(x))$—iff the student is signed up for at least one of the advanced courses, $phil2$ or $cogling$ (rule 3);
- beginner $(introst(x))$—iff the student is signed up for at least one of the introductory courses, $phil1$ or $ling$ (rule 4).

Table 5.1. Exemplary database DB

$\mathcal{EDB}$	$\mathcal{IDB}$
$st(a)$	(1) $phspec(x) \leftrightarrow sig(x, phil1) \wedge sig(x, phil2)$
$st(b)$	(2) $lspec(x) \leftrightarrow sig(x, ling) \wedge sig(x, cogling)$
$st(c)$	(3) $advancedst(x) \leftrightarrow sig(x, phil2) \vee sig(x, cogling)$
$st(d)$	(4) $introst(x) \leftrightarrow sig(x, phil1) \vee sig(x, ling)$
$st(e)$	
$sig(a, phil1)$	
$sig(a, phil2)$	
$sig(b, ling)$	
$sig(b, cogling)$	
$sig(c, ling)$	
$sig(d, cogling)$	
$sig(e, ling)$	

When we consider the possible queries that might be asked of the DB the ones about facts are not as interesting to us as the ones considering concepts introduced in the $\mathcal{IDB}$. A question about this type of concept needs to be decomposed and then executed against the $\mathcal{EDB}$. The decomposition and the plan of checking the facts will be done *via* e-scenarios.

The architecture of the proposed system allows us to say that we will have one e-scenario for each concept introduced in the $\mathcal{IDB}$. What is more the method of construction of this type of e-scenario will be universal, because we allow only for the rules with equivalence as the main connective.

Each scenario will be constructed as follows:

1. Take a rule from the $\mathcal{IDB}$.
2. Construe a simple yest-no question from the formula on the left side of the equivalence.
3. Take the rule as the initial declarative premise of the e-scenario.
4. Decompose the initial question using $\mathsf{lm}(?A, A \leftrightarrow B, ?B)$.

5. After this step build an e-scenario for a formula on the right side of the equivalence. Use $\mathsf{Im}(?(A \otimes B), ? \pm |A, B|)$.

Let us now introduce e-scenarios build for the rules from the $\mathcal{IDB}$. For the first rule (1) $phspec(x) \leftrightarrow sig(x, phil1) \wedge sig(x, phil2)$ we will have the following e-scenario:

(E1)

$$?phspec(x)$$
$$phspec(x) \leftrightarrow sig(x, phil1) \wedge sig(x, phil2)$$
$$?(sig(x, phil1) \wedge sig(x, phil2))$$
$$? \pm |sig(x, phil1), sig(x, phil2)|$$
$$?sig(x, phil1)$$

$$sig(x, phil1) \qquad\qquad \neg sig(x, phil1)$$
$$?sig(x, phil2) \qquad \neg(sig(x, phil1) \wedge sig(x, phil2))$$
$$\qquad\qquad\qquad\qquad \neg phspec(x)$$

$$sig(x, phil2) \qquad\qquad \neg sig(x, phil2)$$
$$sig(x, phil1) \wedge sig(x, phil2) \quad \neg(sig(x, phil1) \wedge sig(x, phil2))$$
$$philspec(x) \qquad\qquad \neg phspec(x)$$

The e-scenario for the second rule: (2) $lspec(x) \leftrightarrow sig(x, ling) \wedge sig(x, cogling)$ will be construed analogically to the one for the rule (1)—I will refer to this e-scenario as (E2). When it comes to rule (3) the e-scenario will be built as follows.

(E3)

$$?advancedst(x)$$
$$advancedst(x) \leftrightarrow sig(x, phil2) \vee sig(x, cogling)$$
$$?(sig(x, phil2) \vee sig(x, cogling))$$
$$? \pm |sig(x, phil2), sig(x, cogling)|$$
$$?sig(x, phil2)$$

$$sig(x, phil2) \qquad\qquad \neg sig(x, phil2)$$
$$sig(x, phil2) \vee sig(x, coling) \qquad ?sig(x, coling)$$
$$advancedst(x)$$

$$sig(x, coling) \qquad\qquad \neg sig(x, coling)$$
$$sig(x, phil2) \vee sig(x, coling) \quad \neg(sig(x, phil2) \vee sig(x, coling))$$
$$advancedst(x) \qquad\qquad \neg advancedst(x)$$

Similarly to the previous case, the e-scenario for the rule (4) will be build analogically to the scenario for the rule (3)—this will be e-scenario (E4).

Let us now consider several examples of interactions that are available for the $COOP_{IEL}$ system.

5.2.1. Affirmative answer

Our aim here is to propose a procedure which will allow for generation of cooperative answers on the basis of an executed path of an e-scenario (i.e. on the basis of e-derivation for the direct answer to the initial question). The following example illustrates such an answer:

Example 6.   USER: Is *a* specialising in philosophy?
  SYSTEM: The answer to your question is yes (*a* is specialising in philosophy).
  SYSTEM: For this solution, the following concepts for the *DB* were used: a student is specialising in philosophy if and only if they are signed up for *phil1* and *phil2*.
  SYSTEM: The following facts were collected while your query was executed against the *DB*: *a* is signed up for *phil1* and *a* is signed up for *phil2*

  The procedure which will generate the answer presented in Example 6 may be roughly described as follows.
  Let us consider the executed path of an e-scenario. Such a path is the e-derivation:

$$\mathbf{s} = \mathbf{s}_1, ..., \mathbf{s}_n$$

where $\mathbf{s}_1$ is the initial question and $\mathbf{s}_n$ is the answer to this question.
  In order to obtain a desired cooperative answer (roughly) the following steps should be taken:

1. Identify queries of this e-derivation and store answers collected that pertain to these queries.
2. Identify the set of initial declarative premises of the e-derivation entangled in obtaining the direct answer and store them.

  Let us remember here the intuition behind the notion of entanglement (introduced in detail in Chapter 2, see page 40). The main intuition here is to identify all the d-wffs from the set of initial premises of a given e-derivation that are relevant for the obtained direct answer to the initial question of this e-derivation.
  In a more detailed manner we may present the procedure as follows.

1. Save $\mathbf{s}_n$ of $\mathbf{s}$ as SOL.
2. Identify queries of $\mathbf{s}$.
3. For every query $\mathbf{s}_k$ ($1 < k < \mathbf{s}_n$) return $\mathbf{s}_{k+1}$, i.e. direct answer to query $\mathbf{s}_k$. Save all the answers in ANS.
4. Identify the set of initial declarative premises of $\mathbf{s}$ entangled in a derivation of the answer to the initial question. Save the set as ENT.
5. Return

  a. The answer to your question is SOL.

b. For this solution, the following concepts for the DB were used: ENT.
c. The following facts were collected while your query was executed against the DB: ANS.

For the presented procedure we should consider one crucial point—namely establishing the set of initial declarative premises of s entangled in the answer to the initial question (ENT). For our case the method of constructing DB ensures that for each rule in the database there will be only one e-scenario relative to this rule. This ensures also that when we consider the e-derivation for a given initial question we can be sure that there will always be only one declarative premise relevant for this question. Further on, since we only use the rules built on the basis of equivalence we ensure that all the answers obtained for the queries will be relevant to the answer to the initial question. This allows us to simplify the procedure of establishing the ENT to the following:

1. Take the initial premise of the e-derivation.
2. Take all the answers to queries of this e-derivation.

One more general observation is in place here. If we consider the more general database schema (e.g. without restrictions on the construction of rules in the IDB part) our cooperative procedure would also work—due to the method of e-scenarios' construction. What would change is the procedure of establishing ENT which would become much more complicated. For general cases the method would consist of the following steps[2]:

1. Establish a set consisting of all the initial premises of the e-derivation and all the answers to the queries of this e-derivation, i.e. $\Theta_\mathbf{s}$.
2. Generate $2^{\Theta_\mathbf{s}}$.
3. Generate the lattice of the elements of $2^{\Theta_\mathbf{s}}$ ordered by inclusion.
4. Starting with $\emptyset$ (from the bottom-up) check the elements of the lattice whether the answer to the initial question is entailed by the element that is being checked.
5. Stop the procedure at the first element that the answer to the initial question is entailed by this element.

Let us consider a slightly simplified e-derivation (2.22a) presented in Chapter 2.3.1.3.

$$\mathbf{s} = ?\{p, q, r\}, s \rightarrow p, \neg s \rightarrow q \lor r, ?s, s, p \tag{5.1}$$

The set $\Theta_\mathbf{s}$ for (5.1), i.e. the set consisting of all the initial premises of the e-derivation of p and all the answers to the queries of this e-derivation would be the following $\Theta_\mathbf{s} = \{s \rightarrow p, \neg s \rightarrow q \lor r, s\}$, while the power set of $\Theta_\mathbf{s}$ is presented below:

---

[2] Such an approach resembles the one presented by Gaasterland et al. (1992) in the context of identifying minimal failing sub-queries for the need of a cooperative answering system.

$$2^{\Theta_s} = \{\emptyset, \{s \to p\}, \{\neg s \to q \vee r\}, \{s\}, \{s \to p, \neg s \to q \vee r\}, \{s \to p, s\},$$
$$\neg s \to q \vee r, s\}, \{s \to p, \neg s \to q \vee r, s\}\}$$

The lattice representation of this power set is presented in Figure 5.2.

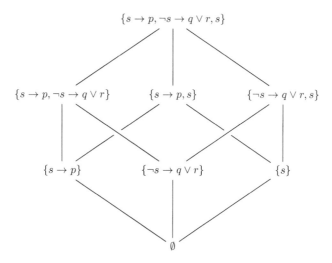

Fig. 5.2. Lattice of the elements of the power set of Θ_s for e-derivation (5.1). Description in the text

     The proposed general method of establishing ENT would be demanding, since the search space might be big for large e-scenarios. However, we can imagine certain heuristics that will make the search easier. For example, if we consider only the information-picking e-scenarios (as it seems to be rather natural in our context) we need not to check the two bottom layers of the lattice. This is due to the characteristic of information-picking e-scenarios the answer to the initial question cannot be entailed by the initial premises themselves nor just the set of answers to queries—it has to be a combination of both.

     Let us now get back to our Example 6. In this case the e-scenario for the user's query would be based on the scheme (E1). After the execution we would obtain the following activated path (e-derivation):

$$?phspec(a) \tag{5.3a}$$

$$phspec(a) \leftrightarrow sig(a, phil1) \wedge sig(a, phil2) \tag{5.3b}$$

$$?(sig(a, phil1) \wedge sig(a, phil2)) \tag{5.3c}$$

$$? \pm |sig(a, phil1), sig(a, phil2)| \tag{5.3d}$$

$$?sig(a, phil1) \tag{5.3e}$$

$$sig(a, phil1) \tag{5.3f}$$

$$?sig(a, phil2) \tag{5.3g}$$

$$sig(a, phil2) \tag{5.3h}$$

$$sig(a, phil1) \wedge sig(a, phil2) \tag{5.3i}$$

$$phlspec(a) \tag{5.3j}$$

For the presented e-derivation the procedure will take formula (5.3j) as SOL, the set ANS will consist of (5.3f) and (5.3h), and (5.3b) will be identified as ENT. Consequently we would obtain the following interaction with DB:

USER: $?phspec(a)$
SYSTEM: The answer to your question is $phspec(a)$.
SYSTEM: For this solution, the following concepts for the DB were used:
$phspec(a) \leftrightarrow sig(a, phil1) \wedge sig(a, phil2)$
SYSTEM: The following facts were collected while your query was executed against the DB:
$sig(a, phil1)$
$sig(a, phil2)$

The explanation contains information about the initial premise of the e-scenario used for the decomposition of the user's question. This piece of information presents our user with the crucial components behind the concept in question (in this case $phspec(a)$). This allows the user to understand how this concept is defined in the database. Further on the user is informed about the facts that were collected while the e-scenario was executed against the database.

In our second example, a user will ask whether student a is also an advanced student, i.e. $?advancedst(a)$. Such a question refers us to the scheme (E3). The activated path of the e-scenario is the following:

$$?advancedst(a) \tag{5.4a}$$

$$advancedst(a) \leftrightarrow sig(a, phil2) \lor sig(a, cogling) \tag{5.4b}$$

$$?(sig(a, phil2) \lor sig(a, cogling)) \tag{5.4c}$$

$$? \pm |sig(a, phil2), sig(a, cogling)| \tag{5.4d}$$

$$?sig(a, phil2) \tag{5.4e}$$

$$sig(a, phil2) \tag{5.4f}$$

$$sig(a, phil2) \lor sig(a, coling) \tag{5.4g}$$

$$advancedst(a) \tag{5.4h}$$

Example 7.   USER: $?advancedst(a)$

SYSTEM: The answer to your question is $advancedst(a)$.

SYSTEM: For this solution, the following concepts for the DB were used: $advancedst(a) \leftrightarrow sig(a, phil2) \lor sig(a, cogling)$

SYSTEM: The following facts were collected while your query was executed against the DB:

$sig(a, phil2)$

5.2.2. Negative answer

When we consider negative answers in the context of our database we may adopt two approaches:

(i)   when a fact is not present in the database it is treated as a negation of this fact,

(ii)   when a fact is not present in the database the search stops—we are not able to establish whether it holds or not.

Let us start with the first approach.

5.2.2.1. Negation as a failure

Consider the following user's question: $?introst(d)$. For the concept of $introst(x)$ the e-scenario scheme (E4) will be used. The following path will be activated:

$$?introst(d) \tag{5.5a}$$

$$introst(d) \leftrightarrow sig(d, phil1) \lor sig(d, ling) \tag{5.5b}$$

$$?(sig(d, phil1) \lor sig(d, ling)) \tag{5.5c}$$

$$? \pm |sig(d, phil1), sig(d, ling)| \tag{5.5d}$$

$$?sig(d, phil1) \tag{5.5e}$$

$$\neg sig(d, phil1) \tag{5.5f}$$

$$?sig(d, ling) \tag{5.5g}$$

$$\neg sig(d, ling) \tag{5.5h}$$

$$\neg(sig(d, phil1) \lor sig(d, ling)) \tag{5.5i}$$

$$\neg introst(d) \tag{5.5j}$$

Observe that formulas (5.5f) and (5.5h) are due to the lack of information in the database. For this situation the interaction would be the following:

Example 8.   USER: $?introst(d)$
    SYSTEM: The answer to your question is $\neg introst(d)$.
    SYSTEM: For this solution, the following concepts for the DB were used: $introst(d) \leftrightarrow sig(d, phil1) \lor sig(d, ling)$
    SYSTEM: The following facts were collected while your query was executed against the DB:
    $\neg sig(d, phil1)$
    $\neg sig(d, ling)$

Let us now consider yet another case. This time we are asking about $?lspec(e)$—scheme (E2) will be in use:

Example 9.   USER: $?lspec(e)$
    SYSTEM: The answer to your question is $\neg lspec(e)$.
    SYSTEM: For this solution, the following concepts for the DB were used: $lspec(e) \leftrightarrow sig(e, ling) \land sig(e, cogling)$
    SYSTEM: The following facts were collected while your query was executed against the DB:
    $sig(e, ling)$
    $\neg sig(e, cogling)$

This is due to the following e-derivation (activated path of (E2)):

$$?lspec(e) \tag{5.6a}$$
$$lspec(e) \leftrightarrow sig(e, ling) \wedge sig(e, cogling) \tag{5.6b}$$
$$?(sig(e, ling) \wedge sig(e, cogling)) \tag{5.6c}$$
$$? \pm |sig(e, ling) \wedge sig(e, cogling)| \tag{5.6d}$$
$$?sig(e, ling) \tag{5.6e}$$
$$sig(e, ling) \tag{5.6f}$$
$$?sig(e, cogling) \tag{5.6g}$$
$$\neg sig(e, cogling) \tag{5.6h}$$
$$\neg(sig(e, ling) \wedge sig(e, cogling)) \tag{5.6i}$$
$$\neg lspec(e) \tag{5.6j}$$

Observe that the formula (5.6f) is present in the database, while (5.6h) cannot be established for our database, thus the answer $\neg sig(e, cogling)$ is introduced.

5.2.2.2. Information gaps

Now we can consider another approach to the lack of information in the database. In this case, when no information is retrieved the execution of an e-scenario stops.

Simple approach

When we take into account the situation in which a user's query fails during its execution against the database due to a lack of information we have to enrich the proposed procedure.

The first change we should introduce is the possibility that the activated path of an e-scenario will not be an e-derivation in the sense presented by Definition 26 (see page 34). However, the only change is that the outcome sequence will differ with respect to the last element—i.e. it will be the sign $\boxtimes$ despite the answer to the initial question. The $\boxtimes$ symbol is used here as an abbreviation for information 'the requested information is not present in the database'. As for the structure and the other elements of such a sequence they will be the same as in the case of an e-derivation.

Our extended procedure starts with an additional step:

0. Check if $s_n = \boxtimes$

   a. if $s_n \neq \boxtimes$ then use the original procedure
   b. if $s_n = \boxtimes$ then follow the next steps

1. Identify the last query. This will be s_{n-1}. Store it as a FQUERY.

2. Identify queries that precede FQUERY in s.
3. For each query s_k $(1 < k < s_{n-1})$ return s_{k+1}, i.e. a direct answer to query s_k. Save all the answers in ANS.
4. Identify the set of initial declarative premises of s entangled in the answer to the initial question. Save the set as ENT.
5. Return

    a. The answer to your question is unknown.
    b. For the solution search the following concepts for the DB were used: ENT.
    c. The following query FQUERY failed.
    d. We have managed to collect the following facts while your query was executed against the DB: ANS.

Let us remember that:

– dQ_s is the set of all the d-wffs of s which are direct answers to queries
– Γ stands for the set of initial declarative premises of s entangled in the direct answer A
– ⊠—the requested information is not present in the database. (In this simple approach at this point the execution of an e-scenario stops.)

If we went back to Example 9 with the question $?lspec(e)$ and scheme (E2) in use, we would obtain the following interaction:

Example 10.   USER: $?lspec(e)$
    SYSTEM: The answer to your question is *unknown.*
    SYSTEM: For this solution, the following concepts for the DB were used: $lspec(e) \leftrightarrow sig(e, ling) \wedge sig(e, cogling)$
    SYSTEM: The following query $?sig(e, cogling)$ failed.
    SYSTEM: We have managed to collect the following facts while your query was executed against the DB:
    $sig(e, ling)$

This is due to the following e-derivation (activated path of (E2)):

$$?lspec(e) \tag{5.7a}$$
$$lspec(e) \leftrightarrow sig(e, ling) \wedge sig(e, cogling) \tag{5.7b}$$
$$?(sig(e, ling) \wedge sig(e, cogling)) \tag{5.7c}$$
$$? \pm |sig(e, ling) \wedge sig(e, cogling)| \tag{5.7d}$$
$$?sig(e, ling) \tag{5.7e}$$
$$sig(e, ling) \tag{5.7f}$$
$$?sig(e, cogling) \tag{5.7g}$$
$$⊠ \tag{5.7h}$$

Observe that formula (5.7h) is introduced due to the failure of searching for an answer to $?sig(e, coling)$ in the database.

Algorithm 1 Cooperative answer generation

Require: $\mathbf{s} = \{s_1, ... s_n\}$
Ensure: cooperative answer
  if $s_n = \boxtimes$ **then**
    FQUERY $\leftarrow s_{n-1}$;
    Find all the answers to queries of $\mathbf{s}$ preceding FQUERY in $\mathbf{s}$.
    ANS $\leftarrow$ the list of answers to the queries preceding FQUERY
    Identify $X_E^{s_n}$
    ENT $\leftarrow X_E^{s_n}$
    print "The answer to your question is unknown."
    print "For the solution search the following concepts for the DB were used:"
    print ENT
    print "The following query:"
    print FQUERY
    print "failed."
    print "We have managed to collect the following facts while your query was executed against the DB:"
    print ANS
  else
    SOL $\leftarrow s_n$
    Find all the answers to queries of $\mathbf{s}$ ($d\mathcal{Q}_{\mathbf{s}}$).
    ANS $\leftarrow d\mathcal{Q}_{\mathbf{s}}$
    Identify $X_E^{s_n}$
    ENT $\leftarrow X_E^{s_n}$
    print "The answer to your question is:"
    print SOL
    print "For this solution, the following concepts for the DB were used:"
    print ENT
    print "The following facts were collected while your query was executed against the DB:"
    print ANS
  end if

Alternative approach

If we would employ the language $\mathcal{L}_{\mathbf{K}3}^?$ introduced in Chapter 2 we can approach the issue of negative answers in an alternative way. Let us remember, that e-scenarios for ternary questions offer three cognitive situations (from the most to the least preferable):

– A 'maximal' cognitive situation is represented by the path that leads to a definite answer to the initial question.
– A 'minimal' one reflects a situation where we did not manage to solve the initial question, but the questioning process ends up with some knowledge gains.
– A 'zero knowledge' situation is represented by a path which ends up without any knowledge gains.

Let us consider a question concerning $?advancedst(d)$. In its ternary form it will be

$$?\{advancedst(d), \neg advancedst(d), \boxminus advancedst(d)\}.$$

For our example we will have to modify the e-scenario scheme (E3)—let us call it (E3'). To save space we will present only the paths relevant to the example.

(E3')

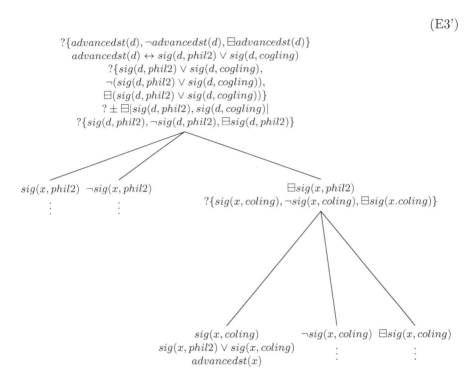

The approach presented here allows for the continuing search for the answer to the user's question despite information gaps. In the analysed case we cannot establish whether it is the case that $sig(d, phil2)$, but we can continue questioning because there is still a path of (E3') which leads to the solution to the initial question. One may notice that the intuition presented here is exactly the one exploited in Chapter 4. In this case we can imagine the following interaction:

Example 11.   USER: $?advancedst(d)$
   SYSTEM: The answer to your question is $advancedst(d)$.
   SYSTEM: For this solution, the following concepts for the DB were used: $adbancedst(d) \leftrightarrow sig(d, phil2) \vee sig(d, cogling)$
   SYSTEM: The following facts were collected while your query was executed against the DB:
   $sig(d, cogling)$

SYSTEM: We were not able to find the following information while your query was executed against the DB:

$sig(d, phil2)$

Now let us consider yet another question. This time it will be a question about $lspec(e)$, which in its ternary version will be formulated as:

$$?\{lspec(e), \neg lspec(e), \boxminus lspec(e)\}.$$

For this case we will introduce the modified e-scenario (E2'). As in the previous example, we will draw only the necessary part of the tree:

(E2')

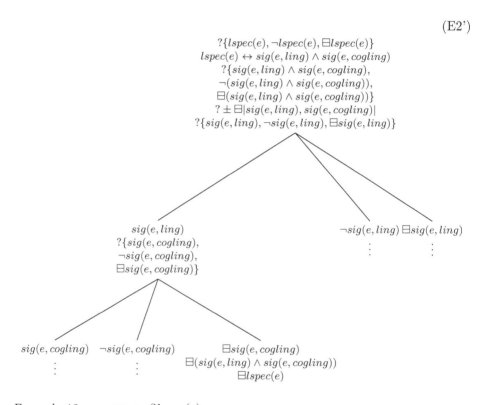

Example 12.   USER: $?lspec(e)$

SYSTEM: The answer to your question is unknown.

SYSTEM: For this solution, the following concepts for the DB were used: $lspec(e) \leftrightarrow sig(e, ling) \wedge sig(e, cogling)$

SYSTEM: The following facts were collected while your query was executed against the DB:

$sig(e, ling)$

SYSTEM: We were not able to find the following information while your query was executed against the DB:
$sig(e, cogling)$

5.2.3. Additional information from a user

We may also imagine that a user gives specific (additional) information together with a question (this information might be gathered during some previous interactions with the DB). An exemplary question of this kind might be:

USER: I <u>know</u> that $sig(a, phil1)$. Is a specialising in philosophy—$?philspec(a)$.

The fact given by the user might be used to shorten the question processing process. In order to express this, we need to introduce a concept of the e-scenarios' *contraction* (see Wiśniewski, 2010). The idea behind this operation on e-scenarios is the following: if I know the answer to one of the auxiliary questions of an e-scenario I may add this answer to the premises of the e-scenario and remove the whole sub-tree generated by the question under consideration (which is the question and its direct answers). This operation is presented schematically in Figure 5.3. More details can be found in (Wiśniewski, 2013b, p. 146–150).

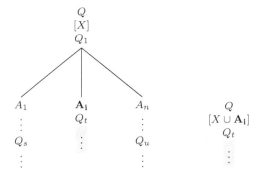

Fig. 5.3. Contraction scheme (see Wiśniewski, 2010)

For our question we will need scheme (E1). For the contraction of this e-scenario the following steps will be needed:

– add the user's statement $(sig(a, phil1))$ to the initial premises
– remove the appropriate sub-tree from the e-scenario, i.e.

$$?sig(a, phil1)$$

$$sig(a, phil1) \quad \neg(sig(a, phil1))$$

The modified e-scenario is presented below:

(E1")

$$?phspec(x)$$
$$phspec(x) \leftrightarrow sig(x, phil1) \wedge sig(x, phil2)$$
$$sig(a, phil1)$$
$$?(sig(x, phil1) \wedge sig(x, phil2))$$
$$? \pm |sig(x, phil1), sig(x, phil2)|$$
$$?sig(x, phil2)$$

$$sig(x, phil2) \qquad\qquad \neg sig(x, phil2)$$
$$sig(x, phil1) \wedge sig(x, phil2) \qquad \neg(sig(x, phil1) \wedge sig(x, phil2))$$
$$philspec(x) \qquad\qquad \neg phspec(x)$$

Execution of this e-scenario against the DB activates the leftmost path, so the answer to the user's question would be the following:

USER: I <u>know</u> that $sig(a, phil1)$.
USER: $?philspec(a)$.
SYSTEM: The answer to your question is $phspec(a)$.
SYSTEM: For this solution, the following concepts for the DB were used:
    $phspec(a) \leftrightarrow sig(a, phil1) \wedge sig(a, phil2)$
SYSTEM: The following fact was provided: $sig(a, phil1)$
SYSTEM: The following facts were collected while your query was executed against the DB:
    $sig(a, phil2)$

We may also imagine that the question asked by a user is enriched with such information, that the question ought not to be executed against the database. This is illustrated by the following example:

USER: I <u>know</u> that $sig(a, phil2)$.
USER: $?advancedst(a)$
SYSTEM: The answer to your question is $advancedst(a)$.
SYSTEM: For this solution, the following concepts for the DB were used:
    $advancedst(a) \leftrightarrow sig(a, phil2) \vee sig(a, cogling)$
SYSTEM: The following fact was provided: $sig(a, phil2)$
SYSTEM: The following facts were collected while your query was executed against the DB: —

In this case the user provided all the necessary information needed to answer the question. However, the user is not aware of the internal database scheme, i.e. that: $advancedst(a) \leftrightarrow sig(a, phil2) \lor sig(a, cogling)$. When we use the contraction operation for this case we obtain only one e-derivation, thus there is no e-scenario to be executed.

$$?advancedst(a) \tag{5.8a}$$

$$advancedst(a) \leftrightarrow sig(a, phil2) \lor sig(a, cogling) \tag{5.8b}$$

$$sig(a, phil2) \tag{5.8c}$$

$$sig(a, phil2) \lor sig(a, coling) \tag{5.8d}$$

$$advancedst(a) \tag{5.8e}$$

Analogical approach might also be used when we would record all the questions asked along with the answers given by the system during one session (and store that information in order to be used in future questions processing). Such stored pieces of information would decrease the number of queries executed against the DB and might also serve for the personalisation of the form of answers provided to the user (cf. Gal 1988, p. 79–81).

5.3. Cooperative posing of questions

A review of the literature reveals that techniques developed in the field of co-operative answering are focused on declarative sentences as reactions to the user's queries. In fact, many authors write simply about answers (as declarative sentences). As such, cooperative answering is well-explored and a number of techniques have been developed in this area of research, such as: evaluation of presuppositions of a query; detection and correction of misconceptions in a query (other than a false presupposition); formulation of intensional answers or generalisation of queries and of responses. The next step should be to extend these techniques with a question-posing capability. Then we would rather consider a *cooperative response* than *cooperative answer*—see (Webber, 1985).

Cooperative interfaces enriched with this functionality will be capable of responding to a user's questions with questions of a kind, that will clarify the user's aims and the details of an answer that he/she expects. By analogy to cooperative answering we may call this functionality *cooperative posing of questions*. Also the aims would be very similar:

- inform a user (in an indirect manner) about the database schema (this will influence his/her future search and should allow us to avoid wrongly formulated questions, and to correct the user's misconceptions);
- providing a user with an answer of a specific form (adjusted to the level of his/her needs—cf. (Gal, 1988));

- personalisation of user's questioning process;
- adjusting the level of generality of answers provided to the user.

Benamara and Saint Dizier (2003) gathered a corpus elaborated from Frequently Asked Questions sections of various web services. Besides well-formed questions the corpus revealed some interesting types of questions like:

- questions including fuzzy terms (like *a cheap country cottage close to the seaside in the Cote d'Azur*);
- incomplete questions (like *When are flights to Toulouse?*);
- questions based on series of examples (like *I am looking for country cottages in mountains similar to Mr. Dupond's cottage*).

These questions were not taken into account in the WEBCOOP development process (because of the early stage of the project at the moment). However, these types of questions asked by users suggests that questions should also be allowed as responses in such systems. A question posing ability would enable us to ask a user for missing information not expressed in his/her question in a natural dialogue manner. The motivation comes from everyday natural language dialogues. As Ginzburg (2010, p. 122) points out:

> Any inspection of corpora, nonetheless, reveals the undiscussed fact that many queries are responded with a query. A large proportion of these are clarification requests (...) But in addition to these, there are query responses whose content directly addresses the question posed (...)

This fact was also noticed by researchers working with databases. Motro (1994, p. 444) writes:

> When presented with questions, the responses of humans often go beyond simple, direct answers. For example, a person asked a question may prefer to answer a related question, or this person may provide additional information that justifies or explains the answer.

5.3.1. When is a question cooperative?

First of all, I must stress that a question might be considered as a cooperative only in a given context, i.e. the context of a question asked by a user, or in the broader context of a user's session with a system. By introducing question posing capabilities into cooperative interfaces we somehow aim at a more general concept of cooperation than the Gricean one (which is much more speaker-centred, cf. Amores and Quesada 2002). In a dialogue context it would be more intuitive to say, that "parties can be seen to be cooperating in dialogue if they obey a common set of (dialogue-) rules and stay within a mutually acknowledged framework" (Reed and Long, 1997, p. 76). This brings us back to the issue of the relevance of a question as a q-specific response as discussed in Chapter 1.1.1.

To clarify this intuition, let us now consider natural language examples of questions which might be treated as posed cooperatively. Let us first look at examples given by Ginzburg (2010, p. 123):

(21) a. A: Who murdered Smith?
          B: Who was in town?

      b. A: Who is going to win the race?
          B: Who is going to participate?

In both examples we may notice that the question is given as a response to the initial question. What is more these response-questions are related to the initial ones. Answers to the questions given in response are somehow connected with the answers to the initial questions. Intuitively, the relation is that the question given as a response somehow restricts the search space for answers to the initial question. In fact it can be shown that e-implication holds in these cases (albeit with some enthymematic declarative premises involved)—cf. page 53.

Let us now take a closer look at the examples taken from the British National Corpus (BNC).

(22) a. CAROL: Right, what do you want for your dinner?
          CHRIS: What do you (pause) suggest?
                  [KBJ, 1936–1937]

      b. AUDREY: Oh! That one's a bit big. Have you got anything a bit
                   smaller than that please?
          UNKNOWN: How much smaller do you want?
                    [KBC, 703–704]

      c. M: They've been done, have they, they haven't been circulated,
               but they're available.
          M: But if everybody would like them?
          SM: Who would like them?
               [KRY, 54–56]

What is interesting, similar relations might be observed between questions being paraphrased by one person. This might be observed especially when we think of the context of learning or questioning someone (see example (7)).

(23) a. IF: The Chairman of the County Council has just got food
                poisoning (pause) Where did he pick it up?
          IF: What restaurant was he in?
          IF: Have you ever inspected the restaurant?
                [KRP, 250–252]

b. UNKNOWN: Question six (pause) okay for anybody who's interested in eating, as we are, *pate de foie gras* is made from
what?

    UNKNOWN: Right we'll be even more specific right, a help for
ya, *pate de foie gras* is made from the liver of what?
[*KDC, 20–21*]

Questions given as responses in the above examples might be treated as
cooperative, because they are closely related to the initial questions and in a
sense they are helpful in getting answers to those initial questions. A cooperative question as a response should establish a search space for the answer to
the initial question which will make it easier to deliver an answer to the initial
question. As we can observe in the examples given, questions as responses ask
for more detailed information (e.g. a list of participants of the race in example
(2) or a suggestion of dishes for the dinner in example (3)). Let us take a closer
look at example (3). We can say that Carol's main aim in this dialogue is to
establish, what Chris would like to eat for dinner. We can also say that both
participants of the dialogue want to reach the decision as soon as possible (to
assure that the dinner will be ready on time). It is rather easy to imagine two
possible reactions to Carol's question. The first one is present in example (3) —
Chris responses with the question: "What do you (pause) suggest?". Probably
an answer given by Carol would be something like: "I would suggest x, y or z.",
or "Well, we may have x or y, but I would prefer z". After an answer like this
it would probably be an easier task for Chris to point out one of the suggested
(and available) dishes. The second of Chris' reactions to Carol's question might
be a simple statement, like "I would like x". We may, however imagine that (at
least in some situations) this answer may lead to some further explications
in the dialogue. As a consequence a delay in establishing a common goal of
the dialogue participants appears. A possible dialogue of this kind is presented
below:

CAROL: Right, what do you want for your dinner?
CHRIS: I'd like x.
CAROL: Oh, I'm sorry, we can't eat x, because we don't have the ingre-
dients to make it...
CHRIS: Let's have y then.
CAROL: Unfortunately it will take to long to prepare y today.

And so on. When we compare those two possible reactions to Carol's question
(i.e. one response with a question and the other one with a statement) it appears
that a response-question might create a way of obtaining the answer to the
initial question easier and shorter. It would be much easier for Chris to make
a choice from an established list, than to guess at what the possible choices
are in the first place. That is why I claim that searching for an answer to
such questions would be easier than for the initial ones. The same might be
observed in examples (6) and (7), when questions are paraphrased in such a

way that they restrict the search area and in consequence enable us to provide easier answer. What is more, when the initial question has a true answer also a question given as a response should have one.

To conclude, we may say that yet again we are aiming at the dependent query responses here. This class of questions constitutes the main intuition behind relevance and cooperativeness in our context. As discussed in details in Chapter 3 we will grasp the dependency of questions with the use of e-implication.

5.3.2. Generating cooperative question-responses by means of erotetic search scenarios

As explained in the previous section, the approach presented here uses the idea of dependency between questions. The direct inspiration is however the class of the so-called FORM questions revealed in the typology of query responses presented in (Łupkowski and Ginzburg, 2013) (see also Chapter 3 of this book). FORM questions address the issue of the *way* the answer to the initial question should be given. In other words, whether the answer to the initial question will be satisfactory to a questioner depends on this kind of question-response.

We may observe this intuition in the following examples:

(24)   A: *Okay then, Hannah, what, what happened in your group?*
       B: **Right, do you want me to go through every point?**
            [*K75, 220–221*]
            [*The way B answers A's question in this case will be dictated by A's answer to question-response—whether or not A wants to know details point by point.*]

(25)   A: [last or full name] you <pause> you,
            you've been a communist yourself?
       B: **Let me give you my family history shall I?**
       A: Oh, oh, if you can do it in a sentence.
       B: I'll do it very quickly.
            [*KJS 245–248*]

The FORM type of question-responses constitute an interesting candidate to use within the domain of cooperative responses. With this response we address directly the question asked and we establish the way the answer to this question should be given. This allows us to produce a better answer to the initial question—which, as a result, will be better suited to a user's needs.

In what follows we will present a procedure of generating such responses, the underlying mechanism of which is based on the dependency relation between questions.

5.3.3. The procedure

As e-scenarios are stored in the cooperative layer between a user and a database, each question of a user might be processed and analysed against these e-scenarios. The layer copes with interactions between a user and a database:

1. First a user's question is analysed in the layer.
2. Then—when it is needed—a question-response is generated.
3. After the user's reaction the next step is executed.
4. On the basis of this interaction the cooperative layer may interact with the database.
5. After the interaction it analyses the data obtained and it may supplement the data with additional explanations useful for the user.

We will consider two types of user's questions: (i) about facts (i.e. concerning the $\mathcal{EDB}$ part) and (ii) about concepts introduced in the $\mathcal{IDB}$ part of the database. In both cases the procedure would be the same. The task will be to find the user's query in e-scenarios stored in the cooperative layer. When a query is found, its position in the e-scenario should be checked. There are two possibilities:

(i)   the user's question is the initial question of the e-scenario (in our example, e.g. questions of the form $?phspec(x)$);
(ii) the user's question is one of the queries of the e-scenario (e.g. questions of the form $?sig(x, phil1)$).

Now—on the basis of this search—we may generate two types of question-response before executing a user's query against the database:

1. *Were you aware that your question depends on the following questions ...? Would you also like to know the answers to them?*

   This question-response allows a user to decide how many details he/she wants to obtain in the answer. This might also be potentially useful for a future search.
2. *Your question influences a higher level question. Will you elaborate on this subject (follow search in this topic)? May I offer a higher level search?*

The procedure boils down to matching a user's question with a question present among the stored e-scenarios. Then generating a question-response is rather straightforward. It should be stressed that it is possible due to the properties of e-scenarios. It is the logical background that facilitates the procedure.

If the user, informed about the dependency relation between his/her question and some other questions, is further interested in a detailed description of connections between the questions, then the program should perform an analysis of such connections and return its results.[3] In our case this step is modelled

---

[3] In the case of really big e-scenarios (with many auxiliary questions) it would be useful to allow a user to decide how many questions influencing the initial question to report.

by returning the whole e-scenario to the user. Algorithm 2 presents a simplified schema of our program's work.

Algorithm 2 User's question analysis and question-response generation

---

Require: question Q to be analysed, e-scenarios (numbered)
Ensure: question-response of an appropriate type
  $n \leftarrow$ the number of e-scenarios in the data base
  for $i = 1$ **to** n **do**
    if question Q is the initial question of e-scenario i **then**
      find the queries of e-scenario i;
      $L \leftarrow$ the list of the queries;
      print "Were you aware that your question depends on the following questions:"
      print L
      print "Are you interested in connections between these questions?"
      if the user answers YES **then**
        return e-scenario i
      end if
    else
      if question Q is one of the queries of e-scenario i **then**
        $Q^* \leftarrow$ the initial question of e-scenario i;
        print "Your question influences a higher level question:"
        print Q^*
        print "Will you elaborate on this subject? Are you interested in connections between these questions?"
        if the user answers YES **then**
          return e-scenario i
        end if
      end if
    end if
  end for

The procedure was proposed in (Łupkowski and Leszczyńska-Jasion, 2014) and has been implemented in Prolog by Dorota Leszczyńska-Jasion. The implementation uses the database concerning the users of a certain computer system presented in Table 5.2. and e-scenarios of the form illustrated in Figure 5.4. The following examples are direct outputs of our implementation (for details see Łupkowski and Leszczyńska-Jasion 2014).[4]

In the first example the user's question is whether a is a local user. This question is recognised as influencing the higher level question (which is the initial question of the e-scenario for the data base in use—see Figure 5.4.) and this fact is reported to the user.

```
?- analyse(?usr(a)).
Your question influences a higher level question: ?locusr(a)
Will you elaborate on this subject?
```

---

[4] The Prolog program with an exemplary database (see Table 5.2.) is available for download from https://intquestpro.wordpress.com/resources/software/

Table 5.2. The database used in the Prolog implementation. Description in the text

$\mathcal{E}DB$	$\mathcal{I}DB$
$usr(a)$	$locusr(x) \to usr(x)$
$usr(b)$	$locusr(x) \to live(x, p)$
$usr(c)$	$usr(x) \wedge live(x, p) \to locusr(x)$
$live(a, p)$	
$live(b, zg)$	
$live(c, p)$	

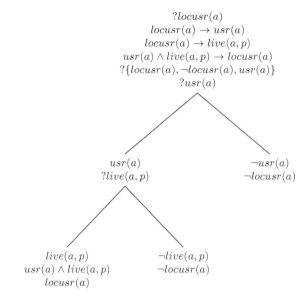

Fig. 5.4. E-scenario for a question 'Is a a local user?'

```
You may search for e-scenario no 1.
true.
```

Another question considering whether a lives in p is also recognised as influencing a higher level question $?locusr(a)$. The system behaviour is analogous to the previous case.

```
?- analyse(?live(a,p)).
Your question influences a higher level question: ?locusr(a)
Will you elaborate on this subject?
You may search for e-scenario no 1.
true.
```

And finally the user poses the question which is recognised as the initial question of the e-scenario. Thus, the system reports questions on which the ini-

tial questions depends (queries retrieved from the e-scenario—see Figure 5.4.):
$?usr(a)$, $?live(a,p)$.

```
?- analyse(?locusr(a)).
Were you aware that your question depends on the following
questions: ?usr(a), ?live(a,p),
You may ask me these questions. You may also search
for e-scenario no 1.
true.
```

What is important, all the data needed to generate question-responses of the
analysed kinds are obtained on the basis of the e-scenarios analysis before their
execution against the database (i.e. the analysis is performed in the cooperative
layer).

We may now consider examples of interactions with IEL_{COOP} including
the presented cooperative features. In these examples we operate on the DB
introduced in Table 5.1.

Example 13. (A question influencing higher level questions, the user uses the
system's suggestion)

USER: $?sig(d, phil1)$
SYSTEM: Your question influences a higher level question $?introst(d)$.
Will you elaborate on this subject? You may search for e-scenario
no 4.
Your question influences a higher level question $?phspec(d)$. Will
you elaborate on this subject? You may search for e-scenario no 1.
USER: $?introst(d)$
SYSTEM: The answer to your question is $\neg introst(d)$.
SYSTEM: For this solution, the following concepts for the DB were
used: $introst(d) \leftrightarrow sig(d, phil1) \lor sig(d, ling)$
SYSTEM: The following facts were collected while your query was
executed against the DB:
$\neg sig(d, phil1)$
$\neg sig(d, ling)$

Example 14. (A question influencing higher level questions, the user does not
use the system's suggestion)

USER: $?sig(a, phil1)$
SYSTEM: Your question influences a higher level question $?introst(d)$.
Will you elaborate on this subject? You may search for e-scenario
no 4.
Your question influences a higher level question $?phspec(d)$. Will
you elaborate on this subject? You may search for e-scenario no 1.
USER: No.

SYSTEM: The answer to your question is $sig(a, phil1)$.

Example 15. (A question depending on lower level questions, the user proceeds with the initial question)

USER: $?advancedst(a)$
SYSTEM: Were you aware that your question depends on the following questions:
$?sig(a, phil2)$
$?sig(a, cogling)$?
You may ask these questions. You may also proceed with $?advancedst(a)$.
USER: Yes.
SYSTEM: The answer to your question is $advancedst(a)$.
SYSTEM: For this solution, the following concepts for the DB were used: $advancedst(a) \leftrightarrow sig(a, phil2) \lor sig(a, cogling)$
SYSTEM: The following facts were collected while your query was executed against the DB:
$sig(a, phil2)$

5.4. Dialogue Logic for cooperative answering

As an integrative approach to the considerations presented in the previous sections I propose a dialogue logic systems $DL(IEL)_{COOP1}$ and $DL(IEL)_{COOP2}$ in the style introduced in details in Chapter 4. The first system is able to generate cooperative responses without the cooperative questioning capacity. $DL(IEL)_{COOP2}$ is extended in such a way that it is able to interact with a user using cooperative questioning. Dialogue logic systems allows us to present, in a clear way, interactions between $COOP_{IEL}$ and a user in a step-by-step dialogue manner as they are governed by the interaction rules.

5.4.1. System $DL(IEL)_{COOP1}$

Let us imagine an interaction between a User and our System (the Cooperative Layer) as a certain dialogue game.

First, let us consider a simple system $DL(IEL)_{COOP1}$, where the System only answers a User's questions (i.e. without the questioning capacity of the System). In this setting User (U) can:

– ask questions.

While System (S) can:

– provide answers;

- provide cooperative explanations, which are:

  - premises used to decompose U's questions,
  - failed queries,
  - collected facts.

For $DL(IEL)_{COOP1}$ we will specify:

1. The taxonomy of locutions.
2. Commitment store rules.
3. Interaction rules.

A dialogue is a k-step finite game between U and S. Each move consists of a locution event performed by one of the players (done one-by-one). The first move is always performed by U, and it is a question.

Each step in a game will be presented as follows:

$$\langle n, \phi, \psi, U \rangle$$

where

n $(1 \leq n \leq k)$ is the number of a step in the dialogue;
ϕ is an agent producing the utterance;
ψ is an addressee;
U is the locution of the agent ϕ.

The following *locutions* are allowed in the $DL(IEL)_{COOP1}$ system:

Categorical statements   including:

1. $A, \neg A, A \wedge B, A \leftrightarrow B, A \leftrightarrow B$,
2. $\boxtimes$ for the unknown answer[5]
3. $FQ(Q_i)$ for a failed query (while searching for answer to Q_i)—identified as FQUERY by the previously presented procedure,
4. $PREM(Q_i)$ premises used to decompose Q_i—ENT in the procedure,
5. $FACTS(Q_i)$ facts gathered while searching for an answer to Q_i—ANS in the procedure.

Questions $?Q_i$. Questions asked by U.

$DL(IEL)_{COOP1}$ consists of the following *interaction rules*:

(In0)   GameStart When $\langle n, U, S, Q_i \rangle$, the game starts.
The rule states that the game between the User and the System is initiated by the User asking a question.

---

[5] $DL(IEL)_{COOP1}$ implements the simplified approach to a lack of information in the DB. The solution using $\mathcal{L}^?_{K3}$ might be incorporated into the system in the way presented in Chapter 4 for systems $DL(IEL)_2$ and $DL(IEL)_{mult}$.

(In1)   Repstat No statement may occur if it is in the commitment store of both participants. This rule prevents pointless repetitions.

(In2)   Q-Response When $\langle n, U, S, Q_i \rangle$, then $\langle n+1, S, U, ans(Q_i) \rangle$
This rule regulates that after the question is asked by the User a system should react by providing an answer to this question (including the situation when the answer is unknown).

(In3)   Q-Explanation
The rule defines the explanations that are attached to the answer to the User's question. The type of explanation differs for the case when the answer is unknown (In3A). For such a case the system reports the failing query (i.e. the one whose execution against the database failed). For the (In3B) one may notice, that this rule tells the system to supplement an answer with premises used to decompose the User's question.

> **(In3A)**   When $\langle n, S, U, d(Q_i) = \boxtimes \rangle$, then
> $\langle n+1, S, U, PREM(Q_i) \rangle$ and
> $\langle n+2, S, U, FQ(Q_i) \rangle$ and
> $\langle n+3, S, U, FACTS(Q_i) \rangle$
> **(In3B)**   When $\langle n, S, U, ans(Q_i) \neq \boxtimes \rangle$, then $\langle n+1, S, U, PREM(Q_i) \rangle$

(In4)   Q-Facts When $\langle n, S, U, PREM(Q_i) \rangle$, then $\langle n+1, S, U, FACTS(Q_i) \rangle$
This rule tells us that the System after returning the premises used to decompose the User's question should supplement them with the collected facts (i.e. answers to queries of the used e-scenario that were executed against the database).

(In5)   EndGame $\langle n, S, U, FACTS(Q_i) \rangle$, then the game ends.
The game ends after the System provides a complete cooperative answer to the User (i.e. an answer to the User's question and additional explanations concerning this answer).

The commitment store (ComSt) rules for $DL(IEL)_{COOP1}$ are rather straightforward.

Rule	Locution	U's ComSt	S's ComSt
(CS1) Q_i		$+Q_i$	$+Q_i$
(CS2) $ans(Q_i)$		$+ans(Q_i)$	$+ans(Q_i)$
(CS3) $PREM(Q_i)$		$+PREM(Q_i)$	$+PREM(Q_i)$
(CS4) $FQ(Q_i)$		$+FQ(Q_i)$	$+FQ(Q_i)$
(CS5) $FACTS(Q_i)$		$+FACTS(Q_i)$	$+FACTS(Q_i)$

Let us now consider examples from the previous sections described with the use of $DL(IEL)_{COOP1}$.

Example 16. (See page 95)
$\langle 1, U, S, ?phspec(a) \rangle$ (In0)
$\langle 2, S, U, phspec(a) \rangle$ (In2)
$\langle 3, S, U, PREM = \{phspec(a) \leftrightarrow sig(a, phil1) \wedge sig(a, phil2)\} \rangle$ (In3B)
$\langle 4, S, U, FACTS = \{sig(a, phil1), sig(a, phil2)\} \rangle$ (In4)
end (In5)

Example 17. (See page 99)
$\langle 1, U, S, ?lspec(e) \rangle$ (In0)
$\langle 2, S, U, \boxtimes \rangle$ (In2)
$\langle 3, S, U, PREM = \{lspec(e) \leftrightarrow sig(e, ling) \wedge sig(e, cogling)\} \rangle$ (In3A)
$\langle 4, S, U, FQ =?sig(e, cogling) \rangle$ (In3A)
$\langle 5, S, U, FACTS = \{sig(e, ling)\} \rangle$ (In4)
end (In5)

5.4.2. System $DL(IEL)_{COOP2}$

In the extension of the simple system $DL(IEL)_{COOP1}$ the User (U) will be
allowed to:

- ask questions;
- provide additional facts (information about U's knowledge).

While the System (S) will be allowed to:

- provide answers;
- provide cooperative explanations

    - premises used to decompose U's questions,
    - failed queries,
    - collected facts.

- ask cooperative questions.

The following *locutions* are allowed in the system $DEL(IEL)_{COOP2}$:

Categorical statements including:

1. A, $\neg A$, $A \wedge B$, $A \leftrightarrow B$, $A \leftrightarrow B$,
2. $\boxtimes$ for the unknown answer,
3. $FQ(Q_i)$ for a failed query (while searching for answer to Q_i)—identified
   as FQUERY by the previously presented procedure,
4. $PREM(Q_i)$ premises used to decompose Q_i—ENT in the procedure,
5. $FACTS(Q_i)$ facts gathered while searching for an answer to Q_i—ANS in
   the procedure.

Questions $?Q_i$. Questions asked by U.

Coop-questions $In(Q_i)$ and $Dep(Q_i)$ (expressing a FORM question about influencing and dependent questions respectively as described on page 110):

$In(Q_i)$ *Your question influences a higher level question. Will you elaborate on this subject (follow search in this topic)? May I offer a higher level search?*

$Dep(Q_i)$ *Were you aware that your question depends on the following questions ...? Would you also like to know the answers to them?*

Coop-questions reactions: yes, no, Q_c , where Q_c is a question pointed out by the user after coop-question.

$DL(IEL)_{COOP2}$ consists of the following *interaction rules*: The rules GameStart, Repstat, Q-Explanation, Q-Facts and EndGame have the same intuitive meaning as in $DL(IEL)_{COOP}$. The rule Q-Response is now replaced with rules (In4) and (In5) to incorporate the effects of interaction with the User after the cooperative question-response is given by the System.

In0   GameStart When $\langle n, U, S, Q_i \rangle$, the game starts.

In1   Repstat No statement may occur if it is in the commitment store of both participants. This rule prevents pointless repetitions.

In2   Coop-Q When $\langle n, U, S, Q_i \rangle$, then $\langle n+1, S, U, In(Q_i) \rangle$ or $\langle n+1, S, U, Dep(Q_i) \rangle$

In3   Coop-Q-Reaction When $\langle n, S, U, In(Q_i) \rangle$ or $\langle n, S, U, Dep(Q_i) \rangle$ then

In3A $\langle U, S, Yes \rangle$
In3B $\langle U, S, No \rangle$
In3C $\langle U, S, Q_c \rangle$

In4   Coop-In(Q)-Resolve

In4A   When $\langle n, U, S, ans(In(Q_i)) = Q_c \rangle$, then
$\langle n+1, S, U, ans(Q_c) \rangle$
$n+2$ Q-Explanation for Q_c
$n+3$ Q-Facts for Q_c
$n+4$ EndGame for Q_c

In4B   When $\langle n, U, S, ans(In(Q_i)) = no \rangle$, then $\langle n+1, S, U, ans(Q_i) \rangle$ and $n+2$ EndGame for Q_i.

In5   Coop-Dep(Q)-Resolve

In5A   When $\langle n, U, S, ans(In(Q_i)) = yes \rangle$, then then $\langle n+1, S, U, ans(Q_i) \rangle$
In5B   When $\langle n, U, S, ans(Dep(Q_i)) = Q_c \rangle$, then $\langle n+1, S, U, ans(Q_c) \rangle$

In6   Q-Explanation

In6A   When $\langle n, S, U, ans(Q_i) = \boxtimes \rangle$, then
$\langle n+1, S, U, PREM(Q_i) \rangle$ and

$\langle n+2, S, U, FQ(Q_i) \rangle$ and
$\langle n+3, S, U, FACTS(Q_i) \rangle$
In6B   When $\langle n, S, U, ans(Q_i) \neq \boxtimes \rangle$, then $\langle n+1, S, U, PREM(Q_i) \rangle$

In7   Q-Facts When $\langle n, S, U, PREM(Q_i) \rangle$, then $\langle n+1, S, U, FACTS(Q_i) \rangle$
In8   EndGame $\langle n, S, U, FACTS(Q_i) \rangle$, then the game ends.

The commitment store (ComSt) rules for $DL(IEL)_{COOP2}$ are straightforward like in the case of the simpler system. Each locution introduced by one of the players goes to both ComSts.

To illustrate the introduced rules let us go back to Example 15 presented on page 114.

USER: $?advancedst(a)$
SYSTEM: Were you aware that your question depends on the following
    questions:
    $?sig(a, phil2)$
    $?sig(a, cogling)$?
    You may ask these questions. You may also proceed with $?advancedst(a)$.
USER: Yes.
SYSTEM: The answer to your question is $advancedst(a)$.
SYSTEM: For this solution, the following concepts for the DB were used:
    $advancedst(a) \leftrightarrow sig(a, phil2) \vee sig(a, cogling)$
SYSTEM: The following facts were collected while your query was executed
    against the DB:
    $sig(a, phil2)$

And now let us represent this interaction within the system $DL(IEL)_{COOP2}$:

Example 18.
$\langle 1, U, S, ?advancedst(a) \rangle$ (In0)
$\langle 2, S, U, Dep(?advancedst(a)) \rangle$ (In2)
$\langle 3, S, U, Yes \rangle$ (In4A)
$\langle 4, S, U, advancedst(a) \rangle$ (In5A)
$\langle 5, S, U, PREM = \{advancedst(a) \leftrightarrow sig(a, phil2) \vee sig(a, cogling)\} \rangle$ (In6B)
$\langle 6, S, U, FACTS = \{sig(a, phil2)\} \rangle$ (In7)
end (In8)

In this example the user asks whether a is an advanced student (step 1). The system recognises this question as depending on questions about a being signed up for *phil2* and *cogling* classes, which is reported to the user (step 2). The user chooses to proceed with the original question $?advancedst(a)$ (step 3), thus the system reports the answer to this question along with additional explanations (steps 4–6). Afterwards the game ends.

Examples 13 and 14 present interactions with the user in a situation when the initial question influences a higher level question. We will observe two reactions of the user after a cooperative question response is given by the system. In the first case, the user will agree on one of the higher level questions proposed by the system and in the second case the user will stay with the original question.

Example 19.
$\langle 1, U, S, ?sig(d, phil1) \rangle$ (In0)
$\langle 2, S, U, In(?sig(d, phil1)) \rangle$ (In2)
$\langle 3, U, S, Q_c = ?introst(d) \rangle$ (In3C)
$\langle 4, S, U, \neg introst(d) \rangle$ (In4A)
$\langle 5, S, U, PREM = \{introst(d) \leftrightarrow sig(d, phil1 \lor sig(d, ling)\} \rangle$ (In4A, In6B)
$\langle 6, S, U, FACTS = \{\neg sig(d, phil), \neg sig(d, ling)\} \rangle$ (In4A, In7)
end (In4B, In8)

Example 20.
$\langle 1, U, S, ?sig(a, phil1) \rangle$ (In0)
$\langle 2, S, U, In(?sig(a, phil1)) \rangle$ (In2)
$\langle 3, U, S, No \rangle$ (In3B)
$\langle 4, S, U, sig(a, phil1) \rangle$ (In4B)
end (In4B, In8)

5.5. Summary

This chapter presents a system rooted in IEL concepts which allows for certain cooperative interaction with a user of a knowledge system. The motivation for the presented system comes from the natural language dialogues discussed in the previous chapters. What is crucial for the system is that the cooperative answering behaviours that are modelled by means of the e-scenarios framework. The system not only analyses users' questions and provides answers enriched with additional explanations. It can also react with questions to users' questions. For this I discuss the idea of cooperative questioning (stemming from the discussions about the dependency of questions and relevance in dialogue). The functionality offered by question-responses is be analogical to the functionality of cooperative answers. Last but not least, involving questions into the cooperative answering process is motivated by the natural language dialogues. As a result, interactions with databases and information systems may become more 'natural' and somehow closer to real-life conversations.

Conclusions and future work

In this book I employ certain logical concepts in my exploration of the wild of natural language dialogues. The underlying motivation comes from the practical (or cognitive) turn in logic as explained in the introduction. My hope at this point of the book is that the results are convincing and interesting to readers of different backgrounds. As may be observed in Chapter 1 dialogue studies are becoming more and more of an interdisciplinary endeavour also involving logical tools. I focus here on only a small piece of the whole complex and dynamic picture that is studied.

What I have managed to achieve in this book is to model erotetic reasoning in dialogues retrieved from natural language corpora. IEL tools have proven useful and user-friendly in dealing with this task. These are not made-up examples, but real dialogues. I show how a dialogue move, leading from one complex question to an auxiliary one, may be justified with the e-implication. I also demonstrate how a dialogue participant's research agenda (or questioning strategy) may be modelled with the use of e-scenarios. Such an approach is, in my opinion, especially interesting when we consider educational contexts. What is more, combining the results of the corpus study concerning query responses with a pragmatical interpretation of e-scenarios appears to be useful for the cooperative answering systems domain. An IEL-based approach to question dependency and questioning strategies may serve as a framework for designing and implementing such cooperative systems. Not only do I present how to generate a cooperative response on the basis of an e-scenario, but I also propose how to add an interaction to theses systems. Thus I introduce a notion of cooperative posing of questions and present a procedure for such system behaviours.

It is worth mentioning that the IEL framework is enriched with a new concept of *entanglement* in the context of an e-derivation. The concept is used to identify all the d-wffs from the set of initial premises of a given e-derivation that are relevant to the obtained direct answer to the initial question of this e-derivation. As such it is a crucial part of the procedures concerning cooperative answering and cooperative posing of questions. As an integrative framework for this book I have proposed dialogic logic systems. These systems allow us to grasp linguistic phenomena and dialogue behaviours in an intuitive and clear way. This approach shares the main intuition with systems presented in (Budzynska et al., 2015; Kacprzak and Budzynska, 2014), namely of using a logic in the background of a formal dialogue system in order to check the correctness of certain dialogue moves. $DL(IEL)_2$ and $DL(IEL)_{mult}$ model verbal behaviours of an information seeking agent using an e-scenario as a questioning strategy in the context of a two- and multi-agent dialogues. I also introduce two systems of this type for a generation of cooperative behaviours. These are $DL(IEL)_{COOP1}$ and $DL(IEL)_{COOP2}$. It is worth noticing that the resulting systems are universal and modular—i.e. they may be fairly easy adapted in order to grasp more aspects of the dynamics of the questioning process (by changing the interaction rules or employing different erotetic logic tools underlying the construction).

As for future work, I reiterate the statement from the Introduction. I have imagined the approach implemented in this book as a form of loop-input for logical concepts concerning erotetic reasoning leading from logic though the empirical domain as a form of a testing field and back to logic again. Using the metaphor of Stenning and Van Lambalgen leading to the wild and back. I believe that this book focuses more on the one arc of this loop-input, namely it says more about erotetic logic concepts in the context of natural language dialogues. Future works should focus more on refining and adapting these concepts in the light of empirical data. It is worth mentioning that the first steps towards this direction are already being taken as discussed in the introduction to Chapter 2.

Summary

The book considers reasoning with questions involved in natural language dialogues. Firstly, I present the overview of linguistic and logical approaches to questions in interaction. The first chapter shows the main points of interests in this area: the issue of the justification for raising a question in a dialogue, the informative content of the act of questioning in interaction with other agents, question relevance and questioning as a strategic process. Considerations of all these issues might be found among the topics of IEL as presented in the second chapter of this book. In what followed the logical ideas were tested and employed for analyses of natural language dialogues retrieved from language corpora. The main result obtained in the third chapter is that erotetic implication and erotetic search scenarios may be successfully used for such a task. The fourth chapter is meant as an integration of linguistic and logical intuitions and concepts. The aim is achieved in the formal dialogue systems presented there. The formalism used in this chapter uses IEL ideas as an underpinning and allows us to grasp the aforementioned issues connected with questions in interaction. The last chapter of this book might be seen as an attempt to take the obtained results one step further. The idea is to involve the logic of questions in solving a practical problem of human–computer interaction related to cooperative answering.

References

Amores, J. G. and Quesada, J. F. (2002). Cooperation and collaboration in natural command language dialogues. In *Proceedings of the sixth workshop on the semantics and pragmatics of dialogue (EDILOG 2002)*, pages 5–11.

Belnap, N. D. (1977). A useful four-valued logic. In Dunn, J. M. and Epstein, G., editors, *Modern uses of multiple-valued logic*, pages 5–37. Springer.

Benamara, F. and Saint-Dizier, P. (2003). Dynamic generation of cooperative natural language responses. In *Proceedings of the EACL Workshop on Natural Language Generation*.

Benamara, F. and Saint Dizier, P. (2003). WEBCOOP: a cooperative question-answering system on the web. In *Proceedings of the tenth conference on European chapter of the Association for Computational Linguistics (Volume 2)*, pages 63–66. Association for Computational Linguistics.

Budzynska, K., Kacprzak, M., Sawicka, A., and Olena, Y. (2015). *Dynamika dialogow w ujeciu formalnym (Dialogue Dynamics: A Formal Approach)*. IFiS PAN, Warszawa.

Cooper, R. (2005). Austinian truth in Martin-Löf type theory. *Research on Language and Computation*, 3(4):333–362.

Cooper, R. (2012). Type theory and semantics in flux. In Kempson, R., Asher, N., and Fernando, T., editors, *Handbook of the Philosophy of Science*, volume 14: Philosophy of Linguistics. Elsevier, Amsterdam.

Cooper, R. and Ginzburg, J. (2015). Type Theory with Records for NL Semantics. In Fox, C. and Lappin, S., editors, *Handbook of Contemporary Semantic Theory, Second Edition*, Oxford. Blackwell.

Gaasterland, T., Godfrey, P., and Minker, J. (1992). An overview of cooperative answering. *Journal of Intelligent Information Systems*, 1:123–157.

Gal, A. (1988). *Cooperative Responses in Deductive Databases*. PhD thesis, University of Maryland, Department of Computer Science.

Genot, E. (2010). The best of all possible worlds: Where interrogative games meet research agendas. *Belief Revision meets Philosophy of Science*, pages 225–252.

Genot, E. J. (2009a). Extensive questions. In *Logic and Its Applications*, pages 131–145. Springer.

Genot, E. J. (2009b). The game of inquiry: the interrogative approach to inquiry and belief revision theory. *Synthese*, 171(2):271–289.

Genot, E. J. and Jacot, J. (2012). How can questions be informative before they are answered? Strategic information in interrogative games. *Episteme*, 9(02):189–204.

Gierasimczuk, N., Maas, H. L. J., and Raijmakers, M. E. J. (2013). An analytic tableaux model for deductive mastermind empirically tested with a massively used online learning system. *Journal of Logic, Language and Information*, 22(3):297–314.

Ginzburg, J. (2010). Relevance for dialogue. In Łupkowski, P. and Purver, M., editors, *Aspects of Semantics and Pragmatics of Dialogue. SemDial 2010, 14th Workshop on the Semantics and Pragmatics of Dialogue*, pages 121–129. Polish Society for Cognitive Science, Poznań.

Ginzburg, J. (2011). Questions: Logic and Interactions. In van Benthem, J. and ter Meulen, A., editors, *Handbook of Logic and Language (Second Edition)*, pages 1133–1146. Elsevier, London.

Ginzburg, J. (2012). *The Interactive Stance: Meaning for Conversation*. Oxford University Press, Oxford.

Ginzburg, J. (2016). The semantics of dialogue. In Aloni, M. and Dekker, P., editors, *The Cambridge Handbook of Formal Semantics*. Cambridge University Press.

Ginzburg, J., Cooper, R., and Fernando, T. (2014). Propositions, questions, and adjectives: a rich type theoretic approach. In *Proceedings of the EACL 2014 Workshop on Type Theory and Natural Language Semantics (TTNLS)*, pages 89–96, Gothenburg, Sweden. Association for Computational Linguistics.

Girle, R. A. (1997). Belief sets and commitment stores. In *Proceedings of the Argumentation Conference*, Brock University, Ontario.

Godfrey, P. (1997). Minimization in cooperative response to failing database queries. *International Journal of Cooperative Information Systems*, 6(2):95–149.

Godfrey, P., Minker, J., and Novik, L. (1994). *Applications of Databases: First International Conference, ADB-94 Vadstena, Sweden, June 21–23, 1994 Proceedings*, chapter An architecture for a cooperative database system, pages 3–24. Springer, Berlin/Heidelberg.

Graesser, A. C. and Person, N. K. (1994). Question asking during tutoring. *American educational research journal*, 31(1):104–137.

Graesser, A. C., Person, N. K., and Huber, J. D. (1992). Mechanisms that generate questions. In Lauer, T. E., Peacock, E., and Graesser, A. C., editors, *Questions and information systems*, pages 167–187. Lawrence Erlbaum Associates, Hillsdale.

Grice, H. P. (1975). Logic and conversation. In Cole, P. and Morgan, J., editors, *Syntax and Semantics*, pages 41–58. Academic Press, New York.

Grobler, A. (2012). Fifth part of the definition of knowledge. *Philosophica*, 86:33–50.

Groenendijk, J. (2009). Inquisitive semantics: Two possibilities for disjunction. In Bosch, P., Gabelaia, D., and Lang, J., editors, *Logic, Language, and Computation*, volume 5422 of *Lecture Notes in Computer Science*, pages 80–94. Springer, Berlin/Heidelberg.

Groenendijk, J. and Roelofsen, F. (2009). Inquisitive semantics and pragmatics. In Larrazabal, J. and Zubeldia, L., editors, *Meaning, Content and Argument. Proceedings of the ILCLI International Workshop on Semantics, Pragmatics and Rhetoric*, pages 41–72. University of the Basque Country Publication Service.

Groenendijk, J. and Roelofsen, F. (2011). Compliance. In Lecomte, A. and Tronçon, S., editors, *Ludics, Dialogue and Interaction*, pages 161–173. Springer-Verlag, Berlin/Heidelberg.

Hamami, Y. (2015). The interrogative model of inquiry meets dynamic epistemic logics. *Synthese*, 192(6):1609–1642.

Hamami, Y. and Roelofsen, F. (2015). Logics of questions. *Synthese*, 192(6):1581–1584.

Hamblin, C. L. (1958). Questions. *The Australasian Journal of Philosophy*, 36:159–168.

Hansson, S. O. (2014). Logic of belief revision. In Zalta, E. N., editor, *The Stanford Encyclopedia of Philosophy*. Stanford University, winter 2014 edition.

Harrah, D. (2002). The Logic of Questions. In Gabbay, D. and Guenthner, F., editors, *Handbook of Philosophical Logic. Second Edition*, pages 1–60. Kluwer, Dordrecht/Boston/London.

Hintikka, J. (1999). *Inquiry as Inquiry: A Logic of Scientific Discovery*, chapter Interrogative Logic as a General Theory of Reasoning, pages 47–90. Springer Netherlands, Dordrecht.

Kacprzak, M. and Budzynska, K. (2014). Strategies in dialogues: A game-theoretic approach. *Computational Models of Argument: Proceedings of COMMA 2014*, 266:333.

Komosinski, M., Kups, A., Leszczyńska-Jasion, D., and Urbański, M. (2014). Identifying efficient abductive hypotheses using multicriteria dominance relation. *ACM Trans. Comput. Logic*, 15(4):28:1–28:20.

Kowtko, J. C. and Price, P. J. (1989). Data collection and analysis in the air travel planning domain. In *Proceedings of the Workshop on Speech and Natural Language*, HLT'89, pages 119–125, Stroudsburg, PA, USA. Association for Computational Linguistics.

Larsson, S. (2002). *Issue based Dialogue Management*. PhD thesis, Gothenburg University.

Larsson, S. and Traum, D. (2003). The information state approach to dialogue management. In van Kuppevelt, J. and Smith, R., editors, *Advances in Discourse and Dialogue*. Kluwer.

Lesniewski, P. and Wisniewski, A. (2001). Reducibility of questions to sets of questions: Some feasibility results. *Logique et analyse*, 44(173-175):93–111.

Leszczyńska, D. (2004). Socratic proofs for some normal modal propositional logics. *Logique et Analyse*, 47(185–188):259–285.

Leszczyńska, D. (2007). *The method of Socratic proofs for normal modal propositional logics*. Adam Mickiewicz University Press.

Leszczyńska-Jasion, D. (2013). Erotetic search scenarios as families of sequences and erotetic search scenarios as trees: two different, yet equal accounts. research report no 1(1)/2013. Technical report, Adam Mickiewicz University, Poznań. https://intquestpro.wordpress.com/resources/reports/.

Leszczyńska-Jasion, D. and Łupkowski, P. (2016). Erotetic search scenarios and three-valued logic. *Journal of Logic, Language and Information*, 25(1):51–76.

Łupkowski, P. (2010). *Test Turinga. Perspektywa sędziego*. Umysł. Prace z filozofii i kognitywistyki. Wydawnictwo Naukowe UAM, Poznań.

Łupkowski, P. (2011a). A Formal Approach to Exploring the Interrogator's Perspective in the Turing Test. *Logic and Logical Philosophy*, 20(1/2):139–158.

Łupkowski, P. (2011b). Human computation—how people solve difficult AI problems (having fun doing it). *Homo Ludens*, 3(1):81–94.

Łupkowski, P. (2015). Question dependency in terms of compliance and erotetic implication. *Logic and Logical Philosophy*, 24(3):357–376.

Łupkowski, P. and Ginzburg, J. (2013). A corpus-based taxonomy of question responses. In *Proceedings of the 10th International Conference on Computational Semantics (IWCS 2013)*, pages 354–361, Potsdam, Germany. Association for Computational Linguistics.

Łupkowski, P., Ignaszak, O., and Wietrzycka, P. (2015). Modelowanie interrogacyjnego rozwiązywania problemów w środowisku gry questgen. *Studia metodologiczne*, 34:239–254.

Łupkowski, P. and Leszczyńska-Jasion, D. (2014). Generating cooperative question-responses by means of erotetic search scenarios. *Logic and Logical Philosophy*, 24(1):61–78.

Łupkowski, P. and Wietrzycka, P. (2015). Gamification for question processing research—the questgen game. *Homo Ludens*, 7(1). in print.

Łupkowski, P. and Wiśniewski, A. (2011). Turing Interrogative Games. *Minds and Machines*, 21(3):435–448.

MacWhinney, B. (2000). *The CHILDES Project: Tools for analyzing talk*. Lawrence Erlbaum Associates, Mahwah, NJ, third edition edition.

McBurney, P. and Parsons, S. (2002). Games that agents play: A formal framework for dialogues between autonomous agents. *Journal of Logic, Language and Information*, 11(3):315–334.

Moradlou, S. and Ginzburg, J. (2014). Learning to Understand Questions. In Muller, P. and Rieser, V., editors, *Proceedings of SemDial 2014 (DialWatt)*, pages 116–124, Edinburgh.

Motro, A. (1994). Intensional answers to database queries. *IEEE Transactions on Knowledge and Data Engineering*, 6(3):444–454.

Noelle, D. C., Dale, R., Warlaumont, A. S., Yoshimi, J., Matlock, T., Jennings, C. D., and Maglio, P. P., editors (2015). *Exploring the processing costs of the exactly and at least readings of bare numerals with event-related brain potentials*, Austin, TX. Cognitive Science Society.

Olney, A. M., Graesser, A. C., and Person, N. K. (2012). Question generation from concept maps. *Dialogue and Discourse*, 3(2):75–99.

Olsson, E. J. and Westlund, D. (2006). On the role of the research agenda in epistemic change. *Erkenntnis*, 65(2):165–183.

Peliš, M. (2010). Set of answers methodology in erotetic epistemic logic. *Acta Universitatis Carolinae Philosophica et Historica*, IX(2):61–74.

Peliš, M. (2011). *Logic of Questions*. PhD thesis, Charles University in Prague.

Peliš, M. (2016). *Inferences with Ignorance: Logics of Questions (Inferential Erotetic Logic & Erotetic Epistemic Logic)*. Karolinum, Praha.

Peliš, M. and Majer, O. (2010). Logic of questions from the viewpoint of dynamic epistemic logic. In Peliš, M., editor, *The Logica Yearbook 2009*, pages 157–172. College Publications, London.

Peliš, M. and Majer, O. (2011). Logic of questions and public announcements. In Bezhanishvili, N., Löbner, S., Schwabe, K., and Spada, L., editors, *Eighth International Tbilisi Symposium on Logic, Language and Computation 2009, Lecture Notes in Computer Science*, pages 145–157. Springer.

Plaza, J. (2007). Logics of public communications. *Synthese*, 158(2):165–179.

Purver, M. (2006). Clarie: Handling clarification requests in a dialogue system. *Research on Language & Computation*, 4(2):259–288.

Reed, C. A. and Long, D. P. (1997). Collaboration, cooperation and dialogue classification. In Jokinen, K., editor, *Working Notes of the IJCAI97 Workshop on Collaboration, Cooperation and Conflict in Dialogue Systems*, pages 73–78.

Rosé, C. P., DiEugenio, B., and Moore, J. (1999). A dialogue-based tutoring system for basic electricity and electronics. In Lajoie, S. P. and Vivet, M., editors, *Artificial intelligence in education*, pages 759–761. IOS, Amsterdam.

Sedlár, I. and Šefránek, J. (2014). Logic and cognitive science. In Kvasnička, V., Pospíchal, J., Návrat, P., Chalupa, D., and Clementis, L., editors, *Artificial Intelligence and Cognitive Science IV*, pages 219–236. Slovak University of Technology Press, Bratislava.

Shoesmith, D. J. and Smiley, T. J. (1978). *Multiple-conclusion Logic*. Cambridge UP, Cambridge.

Stenning, K. and Van Lambalgen, M. (2008). *Human reasoning and cognitive science*. MIT Press.

Švarný, P., Majer, O., and Peliš, M. (2013). Erotetic epistemic logic in private communication protocol. *The Logica Yearbook*, pages 223–238.

Szabolcsi, A. (2015). What do quantifier particles do? *Linguistics and Philosophy*, 38(2):159–204.

Szałas, A. (2013). How an agent might think. *Logic Journal of IGPL*, 21(3):515–535.

Urbański, M. (2001a). Remarks on Synthetic Tableaux for Classical Propositional Calculus. *Bulletin of the Section of Logic*, 30(4):194–204.

Urbański, M. (2001b). Synthetic tableaux and erotetic search scenarios: Extension and extraction. *Logique et Analyse*, 173-174-175:69–91.

Urbański, M. (2002). Synthetic Tableaux for Łukasiewicz's Calculus Ł3. *Logique et Analyse*, 177-178:155–173.

Urbański, M. (2005). Tableaux, abduction and truthlikeness. Technical report, Adam Mickiewicz University, Poznań. http://mu.edu.pl/.

Urbański, M. (2011). Logic and cognition: two faces of psychologism. *Logic and Logical Philosophy*, 20:175–185.

Urbański, M. and Łupkowski, P. (2010a). Erotetic search scenarios: Revealing interrogator's hidden agenda. In Łupkowski, P. and Purver, M., editors, *Aspects of Semantics and Pragmatics of Dialogue. SemDial 2010, 14th Workshop on the Semantics and Pragmatics of Dialogue*, pages 67–74. Polish Society for Cognitive Science, Poznań.

Urbański, M. and Łupkowski, P. (2010b). Erotetic search scenarios: Revealing interrogator's hidden agenda. In Łupkowski, P. and Purver, M., editors, *Aspects of Semantics and Pragmatics of Dialogue. SemDial 2010, 14th Workshop on the Semantics and Pragmatics of Dialogue*, pages 67–74. Polish Society for Cognitive Science, Poznań.

Urbański, M., Paluszkiewicz, K., and Urbańska, J. (2014). Deductive reasoning and learning: A cross curricular study. research report no 2(4)2014. Technical report, Adam Mickiewicz University, Poznań. https://intquestpro.wordpress.com/resources/reports/.

Urbański, M., Paluszkiewicz, K., and Urbańska, J. (2016a). Erotetic problem solving: From real data to formal models. an analysis of solutions to erotetic reasoning test task. In Paglieri, F., editor, *The Psychology of Argument: Cognitive Approaches to Argumentation and Persuasion College Publications*. College Publications, London.

Urbański, M., Paluszkiewicz, K., and Urbańska, J. (2016b). Erotetic Problem Solving: From Real Data to Formal Models. An Analysis of Solutions to Erotetic Reasoning Test Task. In Paglieri, F., editor, *The Psychology of Argument: Cognitive Approaches to Argumentation and Persuasion*. College Publications, London.

Urquhart, A. (2002). Basic many-valued logic. In Gabbay, D. M. and Guenthner, F., editors, *Handbook of Philosophical Logic*, volume 2, pages 249–295. Kluwer AP.

Van Ditmarsch, H., van Der Hoek, W., and Kooi, B. (2007). *Dynamic epistemic logic*, volume 337. Springer Science & Business Media.

van Kuppevelt, J. (1995). Discourse structure, topicality and questioning. *Journal of Linguistics*, 31:109–147.

VanLehn, K., Graesser, A. C., Jackson, G. T., Jordan, P., Olney, A., and Rosé, C. P. (2006). When are tutorial dialogues more effective than reading? *Cognitive Science*, 30:1–60.

Švarný, P., Majer, O., and Peliš, M. (2014). Erotetic epistemic logic in private communication protocol. In Dančák, M. and Punčochář, V., editors, *The Logica Yearbook 2013*, pages 223–237. College Publications, London.

Webber, B. L. (1985). Questions, answers and responses: Interacting with knowledge-base systems. In Brodie, M. and Mylopoulos, J., editors, *On Knowledge Base Management Systems*, pages 365–401. Springer.

Wiśniewski, A. (1994). On the reducibility of questions. *Erkenntnis*, 40(2):265–284.

Wiśniewski, A. (1995). *The Posing of Questions: Logical Foundations of Erotetic Inferences*. Kluwer AP, Dordrecht, Boston, London.

Wiśniewski, A. (1996). The logic of questions as a theory of erotetic arguments. *Synthese*, 109(1):1–25.

Wiśniewski, A. (2001). Questions and inferences. *Logique et Analyse*, 173–175:5–43.

Wiśniewski, A. (2003). Erotetic search scenarios. *Synthese*, 134:389–427.

Wiśniewski, A. (2004a). Erotetic search scenarios, problem-solving, and deduction. *Logique & Analyse*, 185–188:139–166.

Wiśniewski, A. (2004b). Socratic proofs. *Journal of Philosophical Logic*, 33(3):299–326.

Wiśniewski, A. (2010). Erotetic logics. Internal question processing and problem-solving. URL=http://www.staff.amu.edu.pl/~p_lup/aw_pliki/ Unilog.

Wiśniewski, A. (2012). Answering by Means of Questions in View of Inferential Erotetic Logic. In Meheus, J., Weber, E., and Wouters, D., editors, *Logic, Reasoning and Rationality*. Springer.

Wiśniewski, A. (2013a). *Essays in Logical Philosophy*. LIT Verlag, Berlin/Münster/Wien/Zürich/London.

Wiśniewski, A. (2013b). *Questions, Inferences, and Scenarios*. College Publications, London.

Wiśniewski, A. (2015). Semantics of questions. In Lappin, S. and Fox, C., editors, *The Handbook of Contemporary Semantic Theory—second edition*, pages 271–313. Wiley-Blackwell.

Wiśniewski, A. and Leszczyńska-Jasion, D. (2015). Inferential Erotetic Logic Meets Inquisitive Semantics. *Synthese*, 192(6):1585–1608.

Wiśniewski, A. and Shangin, V. (2006). Socratic proofs for quantifiers. *Journal of Philosophical Logic*, 35(2):147–178.

Wiśniewski, A., Vanackere, G., and Leszczyńska, D. (2005). Socratic proofs and paraconsistency: A case study. *Studia Logica*, 80(2):431–466.

Zai, F., Szymanik, J., and Titov, I. (2015). Toward a probabilistic mental logic for the syllogistic fragment of natural language. In Brochhagen, T., Roelofsen, F., and Theiler, N., editors, *Proceedings of the 20th Amsterdam Colloquium*, pages 468–477.

Acronyms and abbreviations

AMEX   The SRI/CMU American Express dialogues
BEE   The Basic Electricity and Electronics Corpus
BNC   The British National Corpus
CHILDES   The Child Language Data Exchange System
ComSt   Commitment Store
CPL   Classical Propositional Logic
CR   clarification request
DB   data base
DP   dependent questions
d-wff   declarative well-formed formula
EDP   Erotetic Decomposition Principle
e-derivation   erotetic derivation
e-implication   erotetic implication
e-scenario   erotetic search scenario
e-wff   erotetic well-formed formula
GPH   Graesser, Person and Huber scheme
FORM   questions considering the way of answering the initial question
IEL   Inferential Erotetic Logic
iff   if and only if
INQ   inquisitive semantics
KoS   KoS is a toponym—the name of an island in the Dodecanese archipelago—bearing a loose connection to *conversation oriented semantics*
MiES   Minimal Erotetic Semantics
mc-entailment   multi-conclusion entailment
NLQ   natural language question
QUD   question under discussion
q-response   query response
rfi   request for information content
TIS   Total Information State
TTR   Type Theory with Records
WEBCOOP   a cooperative question-answering system on the web
wff   well-formed formula

Index

www.ingramcontent.com/pod-product-compliance
Lightning Source LLC
LaVergne TN
LVHW012331060326
832902LV00011B/1822